D0733547

How to Write

SCIENTIFIC AND TECHNICAL

PAPERS

How to Write

SCIENTIFIC AND TECHNICAL

PAPERS

By
SAM F. TRELEASE
Columbia University

The M.I.T. Press
Massachusetts Institute of Technology
Cambridge, Massachusetts, and London, England

The present volume is an outgrowth of two earlier books:
"Preparation of Scientific and Technical Papers" (three editions,
Copyright 1925, 1927, 1936), and "The Scientific Paper, How to
Prepare it, How to Write It" (two editions, Copyright, 1947, 1951).

First M.I.T. Press Paperback Edition, February 1969
Second Paperback Printing, September 1970
Third Paperback Printing, September 1975
Fourth Paperback Printing, April, 1979

ISBN 0 262 70004 2 (paperback)
Library of Congress
Catalog Card Number
58–6803

PREFACE

This manual is intended to meet the practical needs of students and research workers who are preparing illustrated papers or reports on scientific or technical subjects. The student who is confronted with the arduous task of "writing up his data" will find in this book many suggestions that should not only lighten his work but enable him to present his material in a more effective way.

Writing is an essential part of the scientist's profession. The final and in some respects the most important stage in any scientific investigation is the preparation of the results for publication. After a scientific or technical worker has done a good piece of research work, he should present the results to his colleagues in the best possible form. Failure to do so, as may be noted in many cases, may largely discount the value of the work itself. Too often, the technical specialist tends to think that his work is done when the research itself is finished, and to regard the publication as an unnecessary evil and a nuisance. As Charles Darwin expressed it, "a naturalist's life would be a happy one if he had only to observe and never to write."

Few people, in fact, like to write. Certainly few are able to write easily, and those who can compose a good paper in a few hours are indeed rare. The time factor is far more important than the beginner in research is likely to realize. Most candidates for the doctorate should plan to devote about three months to the writing of a dissertation running to the usual length of about forty typewritten pages. The scientist is judged solely by the quality of his final product. No one will criticize him for spending many hours on his manuscript and carrying it through several revisions to make it as nearly perfect as possible.

Unless research workers are willing to learn to write effectively, each scientific or technical institution will need on its staff someone whose duty it is to edit, and, if necessary, ghostwrite publications on work done in the institution. Most of the large industrial organizations and government laboratories engaged in research and development do employ technical writers and editors to aid in the preparation of reports, manuals, technical brochures, journal manuscripts, etc. These writers give valuable editorial service, but they cannot relieve the research worker of the responsibility of preparing an accurate and clear draft of his technical material. And a good scientist or engineer, though welcoming editorial suggestions, wants to do most of the writing himself, in order to present the material in the form that he prefers, to insure accuracy in the presentation, and to have the satisfaction of knowing that the finished work is his own.

Proficiency in writing—like skill in laboratory manipulation—can be gained through study and practice. This essential tool should not be too difficult for the science student to master. Every student who is preparing for scientific or technical research should realize as early as possible that learning to write is a highly important part of his education. In recording the results of laboratory experiments, the student has abundant opportunity for acquiring this skill. To quote A. M. Ramsay (1927):

> The chief purpose of a laboratory training is not to teach the students how to use elaborate instruments or to perform complicated chemical tests. The incalculable gift of the laboratory is its discipline in scientific method, and its training in the importance of logical reasoning and the use of exact language in speaking and in writing. The scientist is always distinguished from the empiricist by

his accuracy in measurement, by his precision in statement, by his honesty in accepting and handling evidence, and by his fairness in presenting it when contesting the opinions of those in opposition to his own views.

The benefit derived from a laboratory course will be increased if each report is carefully reviewed and, if necessary, revised before it is submitted in its final form. The student should strive to write as accurately, clearly, and concisely as possible. To make rapid improvement, he should apply the knowledge he has gained in the study of English composition and should frequently consult a general handbook of composition, as well as a manual dealing specifically with scientific and technical writing. No single factor is more important than daily practice in careful composition. It is also helpful to translate from a foreign language, and to read critically, for their literary style, good books and journal articles.

Several meetings of the departmental seminar for graduate students may profitably be devoted to a discussion of the preparation of scientific results for publication. Each member of the group may report on a phase of the subject in which he is especially interested or competent. To emphasize the points discussed, the reports should be illustrated with examples of good and of poor work selected from the current literature. Techniques used in preparing graphs, drawings, photographs, and lantern slides can be demonstrated by members of the group or by other persons who have acquired skill in these arts. Visits to a photoengraver's shop and a printing plant, which may be arranged for a small party in nearly any city, will enable the group to observe the final steps in the production of printed matter.

This guide to the preparation of papers and reports is intended to be a convenient aid to students and others engaged in scientific or technical work. It has served as a style manual for theses, dissertations, journal articles, and monographs; and it has been used as a supplementary textbook in English courses. This book is the result of a process of development and adaptation. The earlier text, published under the title *The Scientific Paper*, has been revised and expanded, and given a title that indicates more clearly its purposes and contents. It is hoped that the handbook, in its new form, may continue to be of service to students and research workers in scientific and technical fields.

Although, in the main, the directions given in this manual are for the preparation of theses or dissertations, they apply as well to the writing of other types of technical papers and reports in the fields of science, agriculture, engineering, medicine, and industrial research, and to the preparation of manuscripts of a more popular nature on scientific or technical subjects.

Many of the rules included in this book are based on recognized authorities, listed in the bibliography at the end of the volume. Experience in reading students' manuscripts and journal copy has been the guide in selecting the rules and in making the suggestions. Some of the directions are given to achieve uniformity. Although two or more ways may be approved by usage, it is convenient to adopt one form.

No attempt has been made to include rules of grammar and rhetoric, although a few are mentioned as reminders. These subjects are treated so well in general handbooks of English composition that their inclusion here would be superfluous. As preparation for his writing, the advanced graduate student will do well to give himself a month's refresher course in composition, based on one of these books. Details that may have seemed dull and unimpor-

tant acquire new interest and value when they are about
to be put to practical use in a piece of serious writing. It
is taken for granted that every writer will have on his desk
the following books for ready consultation:

Webster's New Collegiate Dictionary
The Concise Oxford Dictionary (For its many examples of cor-
rect use of words and phrases)
*Soule's Dictionary of English Synonyms, Roget's Thesaurus in
Dictionary Form,* or *Webster's Dictionary of Synonyms*
The Perrin Writer's Guide and Index to English, or other good
handbook of English composition.

In general, the style herein suggested conforms to that
of the Waverly Press, a printing house that has developed
a high standard in scientific and technical publications.

I am indebted to Prof. Ronald A. Fisher and Messrs.
Oliver & Boyd Limited, Edinburgh, for permission to re-
print table IV from their book *Statistical Methods for
Research Workers.*

Valuable suggestions and help have come from Prof.
Frederick E. Croxton, Dr. F. E. Denny, Prof. Thomas P.
Fleming, Miss Amy L. Hepburn, Mr. John W. McFarlane,
Prof. Edwin B. Matzke, Prof. Francis B. O'Leary, Prof.
Frank G. Lier, Prof. Francis J. Ryan, Miss Margaret C.
Shields, Prof. Howard W. Vreeland, and Mr. Thomas
W. Yerzley. Special thanks are due Helen M. Trelease
for advice and help during the revision of the book.

CONTENTS

Chapter 1

THE RESEARCH PROBLEM

CHOOSING A RESEARCH PROBLEM

In choosing a research problem, special knowledge of a particular field of science is indispensable. The selection of a problem requires study, thought, and planning—guided by all the imagination, originality, and critical judgment at the command of the investigator.

Suggestions on choosing a problem and on conducting research will be found in Beveridge's (1951) *The Art of Scientific Investigation* and Wilson's (1952) *An Introduction to Scientific Research*. The first emphasizes the qualitative and the second the quantitative aspects of science. Both are so interesting and stimulating that they may be read through in a few sessions, and studied in detail later.

Many scientists find it helpful to accumulate a list, in the form of a card index, of promising research problems from which selection may be made. It is advantageous to make a tentative analysis of each subject and to indicate briefly the object, scope, general plan of investigation, and probable nature of the results that might be obtained.

The criteria given below should be useful in stimulating search and thought during the preliminary survey of possible research problems. Although the criteria are purposely stated in the form of brief rules, it should be understood that they are to be regarded merely as hints or suggestions, which, though helpful in many cases, are not universally applicable.

1. The problem should deal, usually in a quantitative way, with the relations of natural phenomena to one

1

another—or, more specifically, with the conditions that control observable facts and events. Its results should serve as the basis for prediction and control of the phenomena investigated.

2. The problem should be circumscribed, definite, and specific, but it should have a significant bearing on some broad and general field.

3. The problem should be statable in the form of several alternative hypotheses, each of which may be tested in order.

4. The problem should be amenable to experimental treatment with the knowledge and facilities available, and it should give promise of yielding definite and reliable results within the allotted time.

5. The problem should have as its primary object the obtaining of new facts and conclusions in a specific field, preferably on a topic possessing importance to the field as a whole. Its results should be worthy of publication in a leading journal and should be of sufficient importance to deserve mention in a textbook on the subject.

6. The problem should arouse the curiosity and interest of the investigator, give promise of leading to other interesting and important problems, and prepare the investigator for handling them.

7. The problem should usually test some significant proposition about which there is difference of opinion, or one that has secured acceptance upon logically insufficient grounds. It may test the relative merits of two or more mathematically self-consistent hypotheses by checking quantitative predictions based upon each. Obviously, it should be designed to do more than merely disprove some inherently trivial or highly improbable hypothesis.

8. The problem may well deal with the salient features of some new and little-known relation, rather than with

the details of some more thoroughly analyzed and better-known subject. New theoretical and experimental tools, however, often make research on an old topic rewarding. This is particularly true when new techniques allow a considerably more incisive and critical examination of accepted relations than was previously possible.

9. The problem may concern some significant question that has been relatively neglected.

10. The problem should deal with materials that seem to be best adapted to the proposed experimentation. Among materials that seem equally suitable, preference perhaps should be given to ones widely known or economically important.

USING THE LIBRARY[1]

For the research worker, the library plays an essential role. The planning, conducting, interpreting, and reporting of research are all dependent on knowledge gained from the literature available in the library. This section gives an introduction to the use of the library and draws attention to some of the important works to be found there.

1. The Catalogue. The catalogue is the key to the library collections. The card catalogue—in which authors, titles, and subjects are alphabetized in a single file—is most common, although some libraries have their catalogues in book form. Large libraries often have separate catalogues for some materials, especially those not located in the main library. Each branch library—for biology, chemistry, engineering, geology, law, medicine, physics, etc.—is likely to have its separate catalogue. Dissertations and manuscripts are usually filed by themselves. It is well,

---

[1] Prof. Francis B. O'Leary, Librarian, Institute of Technology, University of Minnesota, helped in the preparation of this section.

therefore, to inquire at the reference desk in beginning a search for the literature on your subject.

Before consulting the card catalogue, it is advantageous to find out what special rules have been followed in filing the cards. For example, if the cards have been alphabetized in the logical order, *word by word, New York* comes before *Newfoundland;* but if they have been filed *letter by letter, Newfoundland* comes before *New York.* Names beginning with *Mc* and *M'* are often, but not always, filed as if spelled *Mac.* Libraries usually do not file under prepositions preceding surnames, as *von* and *de;* but they do file under prepositions and articles, as *du.* Hyphenated names are generally filed under the first part of the surname, as *Paige-Wood, John.* When a woman changes her name by marriage, the name under which she first wrote is commonly used, with a cross reference from the married name. Libraries may file *Müller* under *Muller* or *Mueller.* If in doubt, look in both places.

In locating serials or periodicals—including journals, magazines, and publications of societies, institutions, governments, etc.—the best rules to follow are those which appear in the beginning of the *Union List of Serials in Libraries of the United States and Canada.* To quote:

A serial not published by a society or a public office is entered under the first word, not an article, of the latest form of the title.

A serial published by a society, but having a distinctive title, is entered under the title, with reference from the name of the society.

The journals, transactions, proceedings, etc., of a society are entered under the first word, not an article, of the latest form of the name of the society.

Learned societies and academies of Europe, other than English. with names beginning with an adjective denoting royal privilege are entered under the first word following the adjective (Kaiserlich, Königlich, Reale, Imperiale, etc.).

Serials are often catalogued separately, since they form the largest and most used part of a scientific or technical library.

Abbreviated titles in literature citations are often difficult to decipher. In such cases, one should consult the *World List of Scientific Periodicals* and specialized lists of abbreviations, such as those in W. Artelt's *Periodica Medica, Chemical Abstracts,* or *Biological Abstracts.* The *World List* is arranged alphabetically by the first word of the title, not an article, even though the title contains the name of a society. Thus its arrangement is different from that of the *Union List of Serials,* which is usually followed in filing the names of the periodicals in the catalogues of American libraries.

Government documents present a complexity all their own in the catalogues. It is usually best to inquire at the reference department for assistance in obtaining these materials. The United States issues more publications than any other nation. Documents of the United Nations are becoming increasingly important, especially publications of UNESCO. Up-to-date general indexes to publications of many governments and of the United Nations are available. Special indexes are also issued to government publications in certain fields, such as those of the U. S. Geological Survey. Most document series are treated as serials and shelved accordingly.

The results of research done under government contract are published in technical reports. It is often difficult for the library to locate and obtain these. But information on most of them can be found in special indexes, such as *Nuclear Science Abstracts* and *U. S. Government Research Reports.*

2. *Classifications.* Libraries classify their publications in order to bring together those on a particular subject. The most commonly used classifications are the *Dewey decimal system* and the *Library of Congress system*.

The *Dewey decimal system* is a numerical arrangement whereby a number can be expanded by means of a decimal. For example, *General Science* is designated by the number 500, *Chemistry* by 540, and *Organic Chemistry* by 547. Further divisions make use of a decimal, as 547.1, 547.2, etc. Theoretically there are infinite possibilities for expansion. Under the *class number* a series of letters and numbers is added, called the *Cutter number,* that stands for the author. If the author is *Smith,* you will find *Sm 5* or *Sm 55*. These numbers are decimals. *Sm 551* would stand after *Sm 55* but before *Sm 56*. A book on organic chemistry might have the number:

547.11

Sm 55

The *Library of Congress system* classifies by the use of letters and numbers. The letters indicate the class, and the numbers subdivide it. For example, the letter *Q* stands for *Science, QC* for *Physics,* and *QC 391* for *Photometry*. The *Library of Congress system* is preferred by science libraries to the *Dewey decimal system* and is gradually replacing it in such libraries.

Each of these schemes brings together on the shelves the publications that are related in subject matter. The research worker who has access to the shelves should familiarize himself with the arrangement of the books and periodicals. This will enable him to locate many publications quickly and to discover some books that he might otherwise miss.

When requesting a book, it is essential that the *call number*—usually appearing in the upper left-hand corner of the catalogue card—be copied in full. Sometimes an *F* or *Q* appears as a part of the call number, as:

$$547.2 \quad 598.2$$
$$D \; 15 \quad H \; 18$$
$$F \qquad Q$$

These letters indicate a folio or quarto volume that may be shelved apart from the volumes of smaller size. Sometimes a volume number is added:

$$590.6$$
$$Un \; 33$$
$$v. \quad 7$$

This means that the reference is to a certain volume of a series. The call slip should include this volume number and also the title of the series, which usually appears, underlined in red ink, in the center of the catalogue card.

3. Subject Headings. The subject cards in the catalogue are primarily useful for finding textbooks and monographs. The subjects that appear on these cards (usually in red) are selected by specialists. Unfortunately, they may differ from those used in the abstract and index journals. Moreover they are often obsolescent because the library is unable to keep abreast of the rapid changes in terminology that occur in some fields. Most American libraries use headings selected by the Library of Congress, but some modify them and others choose their own.

4. Reference Works. Several general reference works supplement the card catalogue. Chief among these, because of the use of Library of Congress catalogue cards by most libraries, is the *Catalog of Books Represented by Library of Congress Printed Cards, Issued to July 31, 1942,* with its supplement to December 31, 1947. This is continued by the *Library of Congress Author Catalog, 1948–*

1952 and the *Library of Congress Subject Catalog, 1950–1952.* In 1953, the titles were changed to *Library of Congress Catalog, Books: Authors* and *Library of Congress Catalog, Books: Subjects.* In July, 1956, the section *Books: Authors* was changed in title to *National Union Catalog, a Cumulative Author List,* which includes titles reported by American libraries. The *Library of Congress Catalog, Books: Subjects* continues to be published. The Library of Congress began the publication in 1954 of *New Serial Titles, a Union List of Serials Commencing Publication after December 31, 1949.* This serves as a supplement to the *Union List of Serials in Libraries of the United States and Canada.*

The *British Museum, Catalogue of Printed Books* is a comprehensive series covering material through 1899. A revised edition, with the title *General Catalogue of Printed Books,* will begin publication in 1958. This will cover all materials to the end of 1955 in all except the Oriental languages. Equally important is the catalogue of the Bibliothèque Nationale in Paris, which has been completed through *Tendil.* A useful feature is the listing of titles under an author's name, showing in what volume or edition a work may be found.

The current record of books published in the United States is the *Cumulative Book Index,* with its predecessor *U. S. Catalog: Books in Print, 1928.* It is arranged by author and subject and has monthly, quarterly, semiannual, annual, biennial, and quinquennial cumulations. An excellent complement is *Books in Print,* which is the index to the *Publishers' Trade List Annual,* an annual cumulation of the catalogues of the major American publishers. The English equivalent is *Whitaker's Cumulative Book List,* issued quarterly, with annual and quadrennial cumulations.

The *British National Bibliography,* started in 1950, lists weekly, in Dewey classified order and author order, new British books. There are monthly indexes and quarterly and annual cumulations. This bibliography is an indispensable source of reference to British publications. There are also national bibliographies for many other countries.

An essential reference source of the early literature is the *Royal Society of London, Catalogue of Scientific Papers, 1800–1900,* in nineteen volumes. It forms an author index of articles that were published in scientific periodicals, especially those of academies and learned societies. Only three subject indexes have been published—pure mathematics, mechanics, and physics.

Information concerning reference material, bibliographies, dictionaries, encyclopedias, etc., is found, arranged by subjects, in the *Guide to Reference Books* by Constance M. Winchell, published in 1951. Supplements cover new reference material through 1955, and further supplements are planned. An excellent complement is *Science Reference Notes,* published quarterly by the Columbia University Libraries.

5. Aids to Writing Papers. Among the most helpful general guides in the preparation of manuscripts are *A Manual of Style* by the University of Chicago Press, *Words into Type* by Skillin and Gay, and the U. S. Government Printing Office *Style Manual.* Nearly every branch of pure and applied science also has at least one stylebook giving its own rules for publication. An example is the U. S. Geological Survey's *Suggestions to Authors.* Each journal has special rules for style, which are published in the journal or may be obtained from the editor.

6. Affiliated Libraries. The libraries of the various institutions in a large city usually cooperate. Thus the libraries of the New York Botanical Garden and the American Museum of Natural History are affiliated with the Columbia University Libraries. Special libraries have become so numerous that, in order to help their users, they work closely together. The reference department of each library gives information regarding the use of the other libraries, including the hours that they are open.

The Library of Congress maintains the *National Union Catalog*, which records the location of scholarly books and periodicals throughout the United States. There are also a number of regional union catalogues, such as that of the Philadelphia area.

7. *Interlibrary Loans.* A library that does not possess a certain publication may be able to borrow it from another library under a system termed *interlibrary loan*. The research worker should take advantage of this valuable service. In making his request, he should present a complete and accurate citation, and also give its source. But owing to the great demand for such loans, it is often impossible to obtain the needed material except in the form of a photographic copy, described in the section that follows.

8. *Photographic Services.* Reproductions of rare, unusual, or physically deteriorated books, as well as of journal articles, may be secured from most libraries as microfilms or as Photostat copies. A microfilm, usually 35 mm, may be obtained for a nominal sum. Photostat copies, which are usually the same size as the printed page (but may be larger or smaller), are much more expensive than microfilms. The disadvantage of the microfilm (unless expensive, enlarged prints are obtained) is that it must be used in a reading machine of good quality. Information regarding the availability of such a machine in your library may determine whether you wish to obtain a microfilm or a Photostat copy.

9. Map Collections. The use of maps is essential to many types of research. It is important to learn the location of the nearest map collection, the areas and subjects covered, and the rules governing its use. Maps are usually listed in a special catalogue. Various gazetteers and atlases are also available.

GUIDES TO THE LITERATURE[2]

Before making a final selection of a research problem, an investigator needs to learn what other workers have done in his particular field. Aquisition of a broad background knowledge of the field will greatly increase his chances of making a good choice. After having made a tentative selection, he must make a detailed search of the literature to assure himself that his proposed study has not already been made and published.

1. Textbooks and Monographs. In approaching an unfamiliar subject, the research worker should begin with a general source of information, such as a textbook. This will quickly give him a broad view of the subject in its relation to the rest of the field, and will provide him with a summary of current fact and theory in the field. Next, the research worker will do well to consult the most recent monograph that includes the subject of his special interest. A monograph provides a longer and more thorough treatment than is contained in the textbook and indicates phases of the subject that need further investigation.

2. Review Journals. After obtaining a general view of the subject from a textbook and a monograph, the in-

---

[2] Prof. Francis B. O'Leary, Librarian, Institute of Technology, University of Minnesota, helped in the preparation of this section.

vestigator is prepared for understanding and evaluating the more detailed treatment of the subject in an article in a *review journal*. This summarizes knowledge in a particular segment of the field, usually from the beginning of work on that subject to the date of publication. It is selective, touching on the major research in the field and discussing it as fully as possible. The review journal is excellent for rapidly surveying a segment of a larger field.

3. Recent Advances Series, Annual Reviews, and Yearbooks. Newer material, not covered in the review journal, is available in current volumes of the *recent advances series* and the *annual reviews* or *yearbooks*. Together with the review journals, these give the research worker a well-organized general fund of information in his field, and a selective bibliography. They tell him what is currently being done in his research area, and what the reviewers consider to be important.

The *recent advances series* summarizes the progress that has been made in particular fields during a period of several years, since the publication of the preceding volume. It usually covers a shorter period of time than the review journal, and thus can go into greater detail.

The *annual reviews* and *yearbooks* give critical reviews of the literature published on a particular subject during the preceding year, or, in cases of less active phases, during several years. They have chapters or articles by authorities in various fields, and point to important research papers. In using these publications, it is best to begin with the most recent volume and then work back through at least five volumes.

4. Abstract Journals and Index Journals. For newer publications and comprehensive coverage, the *abstract journal* and the *index journal* must be used. The articles in the various types of review periodicals may have missed or neglected some important papers. The investigator who needs to make a comprehensive and detailed search of the literature must rely chiefly on the abstract journal and the index journal. The abstract summarizes the methods used and conclusions reached by the writer of each individual article. It usually gives the research worker enough information to enable him to decide whether or not he needs to read the original article. But since abstracts may not be printed until six to eighteen months after the articles have been published, the more recent articles should be located by consulting an index journal if one is available. This gives a prompt and comprehensive listing of most of the articles on each subject.

5. Current Literature. Very recent publications, which have not been included in abstract and index journals, must be found by examining the current periodicals and books in the library.

6. Original Articles. It is important for the worker to examine all the original papers that he suspects may contain material pertinent to his research. Abstracts may not indicate the full content of the papers, and they, as well as most reviews, have the disadvantage of not presenting evidence in the form of tables and illustrations. Finally, of course, the worker must study critically, in their original, complete form, all the papers that bear directly upon his own research problem.

7. List of Guides to the Literature.[3] The list given below, although by no means complete, contains many of the most useful literature sources in a number of scientific and technical fields. The librarian can give valuable help in locating these and other guides to the literature.

[3] Prepared by Prof. Francis B. O'Leary, Librarian, Institute of Technology, University of Minnesota.

SCIENCE IN GENERAL

Reference Sources.

Encyclopedia Americana.

Encyclopaedia Britannica.

Fleming, T. P.   Guide to the Literature of Science for Use in Courses on Science Literature. 2nd ed. 1957.

Hawkins, R. R.   Scientific, Medical, and Technical Books Published in the United States of America, 1930–1944. Supplements, 1945–1948, 1949–1952.

International Catalogue of Scientific Literature. 17 sections. 1901–1914.

Nature (London). 1869+.

Naturwissenschaften. 1913+.

Royal Society of London. Catalogue of Scientific Papers, 1800–1900.

Sarton, G.   Horus. A Guide to the History of Science. 1952.

Sarton, G.   Introduction to the History of Science. Vol. 1–3. 1927–1948.

Science. 1883+.

Science Reference Notes. 1954+.

Sedgwick, W. T., and H. W. Tyler.   A Short History of Science. Rev. ed. 1939.

Winchell, C. M.   Guide to Reference Books. 7th ed. 1951. Supplements, 1950–1952, 1953–1955. (Quarterly supplements appear in College and Research Libraries, cumulated in printed form every three years.)

Annual Reviews.

Americana Annual. 1923+.

Britannica Book of the Year. 1938+.

Review Journals.

Science Progress. 1906+.

Scientific American. 1845+.

AGRICULTURE AND VETERINARY SCIENCE

Reference Sources.

Agricultural Index. 1916+.

Bailey, L. H.   Standard Cyclopedia of Horticulture. 6 vol. 1914–1917.

Bibliography of Soil Science, Fertilizers and General Agronomy. 1931+.

Index Veterinarius. 1933+.

U. S. Bureau of Animal Industry. Index-Catalogue of Medical and Veterinary Zoology. Rev. ed. 1932–1952. Supplements, Authors A–Q.

U. S. Department of Agriculture. Library. Bibliography of Agriculture. 1942+.

Card Index.

U. S. Department of Agriculture. Card Index [to Its Publications].

Abstract Journals.

Animal Breeding Abstracts. 1933+.

Dairy Science Abstracts. 1939+.

Field Crop Abstracts. 1948+.

Forestry Abstracts. 1939+.

Helminthological Abstracts. 1932+.

Horticultural Abstracts. 1931+.

Plant Breeding Abstracts. 1930+.

Soils and Fertilizers. 1938+.

The Veterinary Bulletin. 1931+.

U. S. Department of Agriculture. Experiment Station Record. 1889–1946.

Annual Reviews.

Advances in Agronomy. 1949+.

Advances in Veterinary Science. 1953+.

ANTHROPOLOGY

Reference Sources.

American Anthropologist. 1888+.

American Antiquity. 1935+.

American Journal of Physical Anthropology. 1918+.

Anthropos. 1906+.

Murdock, G. P. Ethnographic Bibliography of North America. 2nd ed. 1953.

Royal Anthropological Institute of Great Britain and Ireland. Journal. 1871+.

Annual Review.

> Yearbook of Anthropology. 1955+. (Trade edition is titled "Current Anthropology.")
> Yearbook of Physical Anthropology. 1945+.

Review Journal.

> Anthropologischer Anzeiger. 1924–1944.

BOTANY AND ZOOLOGY

Reference Sources.

> Bibliographia Genetica. 1925+.
> Chamberlin, W. J. Entomological Nomenclature and Literature. 2nd ed. 1946.
> Just's Botanisches Jahresbericht. 1873–1939.
> Smith, R. C. Guide to the Literature of the Zoological Sciences. 4th ed. 1955.
> Spector, W. S. Handbook of Biological Data. 1956.
> Zoological Record. 1864+.

Card Indexes.

> Concilium Bibliographicum. 1896–1934.
> Torrey Botanical Club. Index to American Botanical Literature. 1894+.
> Wistar Institute of Anatomy and Biology. Advance Abstract Card Service. 1917+.

Abstract Journals.

> Berichte über die Wissenschaftliche Biologie. 1926+.
> Biological Abstracts. 1926+. (Supersedes Abstracts of Bacteriology, 1917–1926, and Botanical Abstracts, 1918–1926.)
> Botanisches Zentralblatt. 1880–1945.
> British Abstracts AIII: Physiology, Biochemistry, Anatomy. 1945–1953. (Supersedes British Chemical and Physiological Abstracts, 1926–1944.)
> Bulletin Signalétique. Partie 2: Sciences Biologiques. 1940+.
> International Abstracts of Biological Sciences. 1956+. (Supersedes British Abstracts of the Medical Sciences, 1954–1956.)
> Resumptio Genetica. 1924+.

Review of Applied Entomology. 1913+.
Review of Applied Mycology. 1922+.
Zoologischer Bericht. 1922–1944.

Annual Reviews.

Annual Review of Biochemistry. 1932+.
Annual Review of Microbiology. 1947+.
Annual Review of Physiology. 1939+.
Annual Review of Plant Physiology. 1950+.

Recent Advances.

Advances in Biological and Medical Physics. 1948+.
Advances in Food Research. 1948+.
Advances in Genetics. 1947+.
Fortschritte der Botanik. 1931+.
Fortschritte der Zoologie. 1907+.
Progress in Biophysics and Biophysical Chemistry. 1950+.
Survey of Biological Progress. 1949+.

Review Journals.

Biological Reviews (Cambridge Philosophical Society). 1923+.
Botanical Review. 1935+.
Ergebnisse der Biologie. 1926–1943.
Physiological Reviews. 1921+.
Quarterly Review of Biology. 1926+.

CHEMISTRY

Reference Sources.

American Chemical Society. Searching the Chemical Literature.
1951.
Beilstein, F. Handbuch der Organischen Chemie. 4te Auflage.
1918–1938. Supplements I, 1928–1934, and II, 1941+.
Crane, E. J., A. M. Patterson, and E. B. Marr. Guide to the
Literature of Chemistry. 2nd ed. 1957.
Gmelin's Handbuch der Anorganischen Chemie. 8te Auflage.
1924+.
Handbook of Chemistry and Physics. 38th ed. 1956–1957.
Heilbron, I. M., and F. Radt. Elsevier's Encyclopaedia of Or-
ganic Chemistry. 1946+.

Kirk, R. E., and D. F. Othmer. Encyclopedia of Chemical Technology. 15 vol. 1947–1956.

Mellon, M. G. Chemical Publications. 1940.

Mellor, J. W. A Comprehensive Treatise on Inorganic and Theoretical Chemistry. 16 vol. 1922–1937.

Sadler, S. S., et al. Allen's Commercial Organic Analysis. 10 vol. 1923–1933.

Soule, B. A. Library Guide for the Chemist. 1938.

Thorpe, J. F., and M. A. Whitely. Thorpe's Dictionary of Applied Chemistry. 4th ed. 1937+.

Washburn, E. W. International Critical Tables. 8 vol. 1926–1930.

Abstract Journals.

British Abstracts. 1945–1953. (Supersedes British Chemical and Physiological Abstracts. 1926–1944.)

Bulletin Signalétique. Partie I: Mathematique, Astronomie, Physique, Chimie, etc. 1940+.

Chemical Abstracts. 1907+.

Chemisches Zentralblatt. 1830+.

Annual Reviews.

Annual Review of Biochemistry. 1932+.

Annual Review of Physical Chemistry. 1950+.

Chemical Society of London. Annual Reports on the Progress of Chemistry. 1904+.

Society of Chemical Industry, London. Reports on the Progress of Applied Chemistry. 1916+.

Recent Advances.

Advances in Carbohydrate Chemistry. 1945+.

Advances in Catalysis and Related Subjects. 1948+.

Advances in Colloid Chemistry. 1942+.

Advances in Colloid Sciences. 1942+.

Advances in Enzymology and Related Subjects. 1941+.

Advances in Food Research. 1948+.

Advances in Protein Chemistry. 1944+.

Progress in Organic Chemistry. 1952+.

Progress in the Chemistry of Fats. 1953+.

Review Journals.

Chemical Reviews. 1924+.
Chemical Society of London. Quarterly Reviews. 1947+.

ENGINEERING

Reference Sources.

Applied Science and Technology Index. 1958+. (Supersedes
  Industrial Arts Index, 1913–1957.)
Dalton, B. H. Sources of Engineering Information. 1948.
Holmstrom, J. E. Records and Research in Engineering and
  Industrial Science. 3rd ed. 1956.
Industrial Arts Index. 1913–1957.
New York Public Library. New Technical Books. 1915+.
Rimbach, R. How to Find Metallurgical Information. 1936.
Technical Book Review Index. 1935+.

Card Index.

Engineering Index Card Service. 1928+.

Abstract Journals.

Electronic Engineering Master Index. 1925–1949.
Engineering Index. 1884+.
Institution of Civil Engineers, London. Engineering Abstracts.
  1919–1939.
Science Abstracts. Series A: Physics Abstracts, and B: Electrical
  Engineering Abstracts. 1893+.
Technisches Zentralblatt. 1951+.

Recent Advances.

Advances in Applied Mechanics. 1948+.
Advances in Electronics and Electron Physics. 1948+.
Progress in Metal Physics. 1949+.

GEOGRAPHY

Reference Sources.

American Geographical Society. Current Geographical Publications. 1938+.

Bibliographie Cartographique Internationale. 1946+.
Bibliographie Géographique Internationale. 1891+.
The Columbia Lippincott Gazetteer of the World. 1952.
Royal Geographical Society. New Geographical Literature and
    Maps. 1951+.
Webster's Geographical Dictionary. 1949.
Wright, J. K., and E. T. Platt. Aids to Geographical Research.
    2nd ed. 1947.

Review Journal.

Geographical Review. 1916+.

GEOLOGY

Reference Sources.

Bibliographie des Sciences Géologiques. 1923+.
Bibliography and Index of Geology Exclusive of North America.
    1933+.
Bibliography of North American Geology. 1732–1891+.
Geological Society of London. List of Geological Literature. 1894–
    1934.
Hintze, C. Handbuch der Mineralogie. 1897–1923. Ergänzungs-
    band: Neue Mineralien. I, 1936–1938; II, 1954+.
Howell, J. V., and A. I. Levorsen. 1946. Directory of Geological
    Material in North America. (*In* American Association of Pe-
    troleum Geologists, Bulletin 30: 1321–1432.)
Malone, T. F. Compendium of Meteorology. 1951.
Mason, B. The Literature of Geology. 1953.
Palache, C., et al. Dana's System of Mineralogy. 7th ed. 1944–
    1951.
Pearl, R. H. Guide to Geologic Literature. 1951.
Piveteau, J. Traité de Paléontologie. 1952+.
Smithsonian Meteorological Tables. 6th rev. ed. 1951.

Abstract Journals.

Annotated Bibliography of Economic Geology. 1928+.
Geological Abstracts. 1953+.
Geologisches Zentralblatt. 1901–1942.
Geophysical Abstracts. 1929+.
Meteorological Abstracts and Bibliography. 1950+.

Mineralogical Abstracts. 1920+.

Zentralblatt für Geologie und Paläontologie. 1950+.

Zentralblatt für Mineralogie. 1950+. (This and the preceding supersede the Zentralblatt für Mineralogie, Geologie und Paläontologie. 1900–1949.)

Recent Advances.

Advances in Geophysics. 1952+.

Fortschritte der Geologie und Paläontologie. 1923–1943.

Fortschritte der Mineralogie. 1911+.

Physics and Chemistry of the Earth. 1956+.

HISTORICAL, ECONOMIC, POLITICAL, AND SOCIAL SCIENCES

Reference Sources.

American Historical Association. Guide to Historical Literature. 1949.

Burchfield, L. Student's Guide to Materials in Political Science. 1935.

Cambridge Ancient History, Medieval History, Modern History. (Comprehensive series of many volumes.)

Cyclopedia of American Government. 3 vol. 1914.

Encyclopedia of the Social Sciences. 15 vol. 1930–1935.

London Bibliography of the Social Sciences. 4 vol. 1931. Supplements, 1–3, 1929–1950.

New York Times Index. 1913+.

Public Affairs Information Service. Bulletin. 1915+.

Readers' Guide to Periodical Literature. 1900+.

Abstract Journals.

Historical Abstracts. 1955+.

International Political Science Abstracts. 1951+.

Sociological Abstracts. 1952+.

Review Journals.

American Academy of Political and Social Sciences. Annals. 1890+.

American Economics Review. 1911+.

American Historical Review. 1895+.

American Sociological Review. 1936+.

MATHEMATICS

Reference Sources.

> Encyklopädie der Mathematischen Wissenschaften. 6 vol. 1898–1935.
>
> Mathematical Tables Project. Mathematical Tables. 40 vol. 1939–1944.
>
> Parke, N. G. Guide to the Literature of Mathematics and Physics, Including Related Works on Engineering Science. 1947.

Abstract Journals.

> Jahrbuch über die Fortschritte der Mathematik. 1868+.
> Mathematical Reviews. 1940+.
> Zentralblatt für Mathematik und ihre Grenzgebiete. 1931+.

MEDICINE

Reference Sources.

> Current List of Medical Literature. 1941+.
> Index Medicus. 1879–1926.
> Index to Dental Periodical Literature. 1839/1875+.
> Medical Library Association. Handbook of Medical Library Practice. 2nd ed. 1956.
> Oxford Loose-leaf Medicine.
> Oxford Loose-leaf Surgery.
> Postell, W. D. An Introduction to Medical Bibliography. 1945.
> Quarterly Cumulative Index Medicus. 1927+.
> U. S. Armed Forces Medical Library. Index Catalogue of the Library. 1880+. (Will cease with issuance of 4th series, vol. 11: Mi–Mz.)
> U. S. Armed Forces Medical Library Catalog, Authors and Subjects, 1950–1954 (Issued Annually and Cumulated Quinquennially).

Abstract Journals.

> Abstracts of World Medicine and World Surgery. 1947+.
> Berichte über die Gesamte Physiologie und Experimentelle Pharmakologie. 1920+.
> Biological Abstracts. 1926+.

British Abstracts of Medical Sciences. 1954–1956. (Succeeds British Abstracts AIII, 1916–1953; Superseded by International Abstracts of Biological Sciences, 1956+.)

Dental Abstracts. 1956+.

Excerpta Medica. 1947+.

International Abstracts of Surgery. 1913+.

International Medical Digest. 1920+.

Nutrition Abstracts and Reviews. 1931–1932+.

Quarterly Review of Medicine. 1943+.

Tropical Diseases Bulletin. 1912+.

Zentralblatt für die Gesamte Neurologie und Psychiatrie. 1910+.

Zentralblatt für Innere Medizin. 1880+.

Annual Reviews.

Annual Review of Medicine. 1950+.

Annual Review of Physiology. 1939+.

Methods in Medical Research. 1948+.

Yearbook of Dentistry. 1936+.

Yearbook of Drug Therapy. 1933+.

Yearbook of General Surgery. 1901+.

Yearbook of Neurology, Psychiatry and Neurosurgery. 1933+.

Yearbook of Pathology and Clinical Pathology. 1946+.

Yearbook of Pediatrics. 1902+.

Recent Advances.

Advances in Cancer Research. 1942+.

Advances in Internal Medicine. 1942+.

Advances in Pediatrics. 1942+.

Advances in Surgery. 1949+.

Ergebnisse der Physiologie. 1902+.

Recent Advances in Medicine. 1924–1952.

Recent Advances in Pathology. 1932+.

Recent Progress in Hormone Research. 1947+.

Vitamins and Hormones. 1943+.

Review Journals.

Ergebnisse der Innere Medizin und Kinderheilkunde. 1907+.

Fortschritte der Neurologie und Psychiatrie und ihre Grenzgebiete. 1929+.

Medicine. 1922+.

Physiological Reviews. 1921+.

PHYSICS

Reference Sources.

American Institute of Physics. Physics Handbook. 1957.

Eucken, A., and K. L. Wolf. Hand- und Jahrbuch der Chemischen Physik. 1937.

Flügge, S. Handbuch der Physik. 1955+.

Glazebrook, R. A Dictionary of Applied Physics. 5 vol. 1922–1923.

Landolt-Börnstein. Zahlenwerte und Funktionen aus Physik, Chemie, Astronomie, Geophysik und Technik. 6te Auflage. 1950. (Formerly called Physikalisch-Chemische Tabellen.)

Parke, N. G. Guide to the Literature of Mathematics and Physics, Including Related Works on Engineering Science. 1947.

Whitford, R. H. Physics Literature: A Reference Manual. 1954.

Wien, W., and F. Harms. Handbuch der Experimentellen Physik. 28 vol. 1926–1935.

Abstract Journals.

Nuclear Science Abstracts. 1948+.

Physics Abstracts (Section A of Science Abstracts). 1893+.

Physikalische Berichte. 1920+.

Annual Reviews.

Annual Review of Nuclear Science. 1952+.

Journal of Aplied Physics. 1931+.

Physical Society of London. Reports on Progress in Physics. 1934+.

Recent Advances.

Advances in Electronics and Electron Physics. 1948+.

Advances in Physics. 1952+.

Review Journals.

Review of Scientific Instruments. 1930+.

Reviews of Modern Physics. 1929+.

PSYCHOLOGY

Reference Sources.

Contemporary Psychology. 1956+.

Daniel, R. S., and C. M. Louttit. Professional Problems in

Psychology. 1953. (Revision and expansion of Louttit's Handbook of Psychological Literature. 1932.)
Psychological Index. 1894–1935.
Stevens, S. S.   Handbook of Experimental Psychology. 1951.

Abstract Journals.

Biological Abstracts. Section H: Human Biology. 1946–1954.
Bulletin Signalétique. Pt. 3: Sociologie.
Child Development Abstracts. 1927+.
Psychological Abstracts. 1927+.

Annual Reviews.

L'Année Psychologique. 1894+.
Annual Review of Psychology. 1950+.
Mental Measurements Yearbook. 1938+.

Review Journals.

Psychoanalytic Quarterly. 1932+.
Psychological Bulletin. 1904+.

FIRST STEPS IN TREATING DATA

1. Tables. Check all calculations, and put experimental data in the form of tables. (See chapter on "Tables," page 114.) Make all calculations twice, preferably on different days, and, if practicable, by different methods. The second calculation should be made without reference to the first, and on a new page in your notebook. Notes on observational and descriptive work should be arranged and classified.

2. Graphs. Plot your data wherever possible. (See section on "Graphs," page 132, and consult Worthing and Geffner's (1943) *Treatment of Experimental Data*.) In most experimentation, graphs furnish the best means of bringing out relations among data, and should be prepared even if they are not to be published.

It is advisable to keep the data tabulated and plotted

as the experimentation progresses. This will serve to check the accuracy of the work; it will indicate desirable modifications in the plan; and it is likely to suggest valuable new ideas.

3. Written Notes. During the preliminary study of the data, it is advisable to record on paper all the ideas that seem to be worthy of consideration. The mechanical process of recording and filing the notes should suit the convenience of the individual. Some writers like to use a standard size of cards (5 by 8 inches) for all preliminary work on a paper. The cards may be filed in a box, under appropriate headings. Only one topic is put on a card, and this topic is expanded later to make a paragraph. This method allows topics to be added, eliminated, and rearranged whenever necessary. Other writers prefer to use sheets of paper of standard size (8½ by 11 inches), only one topic on a sheet and the sheets filed in folders, large envelopes, or loose-leaf notebooks. (Original observations and measurements are usually recorded in a notebook with permanent pages. A copy should be put in a safe place as soon as possible.)

The writing should be done currently, rather than postponed until shortly before the first draft of the organized paper is due. This will make the preparation of a good first draft relatively easy, by avoiding the necessity of hurried, careless writing at the end. It is best to set aside a regular, uninterrupted half-day period each week for keeping the writing up-to-date.

4. Conclusions. Examine the tables, graphs, and classified notes for relations and conclusions. Ask yourself, "What are the possible explanations of the facts?" If several explanations seem equally probable, do not emphasize *only one*. Consider all logical possibilities. (See section on "Logical Presentation of Ideas," page 44.) Make written notes of tentative conclusions.

If time is available, verify your conclusions by gathering more data or by making special test experiments. Confirm your conclusions, if possible, by evidence from sources that are entirely different in character.

Estimate the probable accuracy of your results by considering the sources of error. Conclusions from your results must be based upon a careful consideration of their accuracy and sufficiency. If you have enough data, use statistical methods to estimate their probable significance.

5. Revision of Conclusions. Refer again to your data to see whether your tentative conclusions are actually justified. Discover in which cases these conclusions apply and in which, if any, they do not. Modify, if necessary, the statement of your conclusions. See also whether they are consistent with established facts or principles pertaining to the subject.

6. Exceptions. Examine the data for exceptions, inconsistencies, and discrepancies. Record the exceptions, and check their values. Try to find out why the expected result was not obtained. Some of the most important scientific discoveries have resulted from apparent exceptions and abnormalities in data. Formulate possible explanations for the exceptions. Study your conclusions again to see how the exceptions modify them.

RELIABILITY AND SIGNIFICANCE OF MEASUREMENTS

This section provides a brief introduction to some aspects of statistical methods. Advice of a specialist should be sought in approaching any but the simplest problem. A general book that should be studied by everyone whose work involves measurements is the one by Wilson (1952).

The following are among the most useful books that deal with statistical methods:

Adams (1955).
Cochran and Cox (1950).
Croxton (1953).
Croxton and Cowden (1939).
Dixon and Massey (1951).
Federer (1955).

Fisher (1951, 1954).
Goulden (1952).
Hald (1952).
Hall (1954).
Walker and Lev (1953).
Youden (1951).

If statistical analysis is to be applied to the results, it is important that the experiments be planned with this in mind. Defects in the design of the experiments may make it difficult or impossible to determine the statistical significance of the results.

The following directions outline a working system for: (a) computing the standard error of the mean of a series of measurements obtained from a sample, (b) ascertaining the significance of the difference between the population mean and the mean of a sample, (c) judging the significance of the difference between two sample means, and (d) estimating the size of an adequate sample.

STANDARD ERROR OF THE MEAN

1. Write the readings in a vertical column. At the bottom of the column, write the sum of the readings; divide this by the number of readings (N) and set down the mean (M).

2. In a second column, put opposite each reading the difference between it and the mean.

3. In a third column, write the square of each difference; and at the bottom of this column, put the sum of these squares (S).

4. Compute the standard error of the mean by taking the square root of the quotient obtained by dividing the

sum of the squares by the product of the number of readings times one less than this number:

$$E_M = [S/(N(N - 1))]^{\frac{1}{2}}$$

5. Write the mean and its standard error in the form $M \pm E_M$. (According to the theory of probabilities, 68 per cent of many similarly determined means, based on large samples, should fall within $\pm E_M$ of the population mean, 95 per cent may be expected to fall within ± 1.96 E_M, and 99 per cent within $\pm 2.58 E_M$.)

SIGNIFICANCE OF THE DIFFERENCE BETWEEN A POPULATION MEAN AND THE MEAN OF A RANDOM SAMPLE

1. Compute n from:

$$n = N - 1$$

where N is the number of readings upon which the mean of the random sample was based.

2. Compute the value of t from:

$$t = (M - m)/E_M$$

where M is the mean of the sample, m is the known or assumed value of the population mean, and E_M is the standard error of the mean of the sample.

3. In table 1,[4] find the *smaller* value of t corresponding to n. If the computed value of t is greater than the value of t found in the table, the difference between the population mean and the mean of the sample may be regarded as significant. If the computed value of t exceeds the larger value of t in the table, the difference is highly significant.

[4] This table is reprinted from table IV of Fisher: *Statistical Methods for Research Workers*, Oliver & Boyd Limited, Edinburgh, by permission of the author and publishers.

A significant difference indicates that the sample was probably not drawn from a population having a mean of m, or that the sample was not drawn at random, or both.

TABLE 1

Values of t at P = 0.05 and P = 0.01 for selected values of n (degrees of freedom) from 4 to infinity*

n	t		n	t	
	$P = 0.05$	$P = 0.01$		$P = 0.05$	$P = 0.01$
4	2.78	4.60	15	2.13	2.95
5	2.57	4.03	16	2.12	2.92
6	2.45	3.71	17	2.11	2.90
7	2.37	3.50	18	2.10	2.88
8	2.31	3.36	19	2.09	2.86
9	2.26	3.25	20	2.09	2.85
10	2.23	3.17	25	2.06	2.79
11	2.20	3.11	30	2.04	2.75
12	2.18	3.06	40	2.02	2.70
13	2.16	3.01	50	2.01	2.68
14	2.15	2.98	Infinity	1.96	2.58

* This table is reprinted from table IV of Fisher: *Statistical Methods for Research Workers*, Oliver & Boyd Limited, Edinburgh, by permission of the author and publishers.

SIGNIFICANCE OF THE DIFFERENCE BETWEEN MEANS

1. Subtract the smaller mean (M_2) from the larger mean (M_1) to obtain the difference (D).

2. Obtain the standard error of the difference between the two means by taking the square root of the sum of the squares of the two standard errors of the means:

$$E_{\mathrm{D}} = [(E_{\mathrm{M_1}})^2 + (E_{\mathrm{M_2}})^2]^{\frac{1}{2}}$$

where $E_{\mathrm{M_1}}$ and $E_{\mathrm{M_2}}$ are the standard errors of the means. This expression for E_{D} should be used only when the two random samples are independent and are about the same size.

3. Write the difference and its standard error in the form $D \pm E_D$.

4. Compute n from:

$$n = (N_1 - 1) + (N_2 - 1)$$

where N_1 and N_2 are the number of readings upon which the means were based.

5. Compute t from $t = D/E_D$. In table 1,[5] find the *smaller* value of t corresponding to n. If the computed ratio is greater than the value of t found in the table, the difference may be considered to be significant. If the ratio exceeds the larger value of t in the table, the difference may be regarded as highly significant. The P (probability) value indicates the probability of obtaining a *plus or minus* difference equal to or greater than that indicated by the value of t.

ADEQUACY OF SAMPLE SIZE

1. An estimate may be made of the size of each of two samples needed in order that a certain percentage difference between the two sample means may be regarded as significant. For simplicity, it is assumed (*a*) that the two populations from which the random samples are drawn have the same degree of variability and (*b*) that the two random samples are independent.

2. Obtain an exploratory random sample of size N (as large as practicable) of one of the two populations, and calculate the mean (M) and the standard error of the mean (E_M).

[5] This table is reprinted from table IV of Fisher: *Statistical Methods for Research Workers*, Oliver & Boyd Limited, Edinburgh, by permission of the author and publishers.

3. Compute the required sample size (N_r) for each of the two samples from the following formula:

$$N_r = 2 \times (2.16)^2 \times N \times (100 \times E_M/M)^2/d^2$$

where 2.16 is a value obtained from a t table and corresponds to a probability of 0.03 for a large sample, and d is the percentage difference desired to be significant.

4. It must be borne in mind that the procedure here outlined can give only a rough estimate of adequate sample size and should not be used for small samples.

Chapter 2

WRITING THE PAPER

OUTLINE OF A SCIENTIFIC PAPER

1. Scientific Writing. A paper on a scientific or technical subject necessarily consists of (*a*) a report of facts, (*b*) an interpretation of facts, or (*c*) a combination of a report and an interpretation. The method of writing is governed by many conditions, including the nature of the subject, the purpose of the article, the characteristics of the writer, and the interests of the probable readers. No set method or arrangement will be suited to all kinds of papers.

It is important that the plan of the composition be made very clear to the reader. The main topics and their subdivisions should be plainly indicated. In this respect scientific writing differs from literary composition. A scientific paper is intended to be studied and used as a reference; it is not merely to be read. Hence literary devices should be subordinated if they interfere with clearness. The plan should be self-evident throughout the composition.

2. General Outline. The outline given below suggests a form that may be used for a wide variety of scientific and technical papers. An examination of the papers published in the journals will show that the majority of them have this general arrangement and sequence of topics. This form of outline is suitable for most papers that report investigations or experiments, and possesses the additional advantage of being familiar to the reader. The outline should be modified sufficiently to adapt it to the special requirements of the article or report that is to be written.

General outline of a scientific paper

TITLE. The title should consist preferably of few words, indicative of the contents that are most emphasized. Great care must be exercised to employ words that contain the elements both of brevity and comprehensiveness and permit easy and accurate indexing.

ABSTRACT. The abstract is a brief condensation of the whole paper.

I. *Introduction.*

   A. Nature of the problem; its state at the beginning of the investigation.
   B. Purpose, scope, and method of the investigation.
   C. Most significant outcome of the investigation; the state of the problem at the end of the investigation.

II. *Materials and methods.*

   A. Description of the equipment and materials employed.
   B. Explanation of the way in which the work was done. (Give sufficient detail to enable a competent worker to repeat your experiments. Emphasize the features that are new.)

III. *Experiments and results.*

   A. Description of the experiments.
   B. Description of the results. (If possible, these should be shown in tables and graphs.)

IV. *Discussion of results.*

   A. Main principles, causal relations, or generalizations that are shown by the results. (Choose one or several main conclusions that your evidence tends to prove.)
   B. Evidence (as shown by the data) for each of the main conclusions.
   C. Exceptions and opposing theories, and explanations of these.
   D. Comparison of your results and interpretations with those of other workers.

Outline of a paper that includes several series of experiments presented separately

TITLE.
ABSTRACT.

I. *General introduction.*

II. *General materials and methods.*

III. *Descriptive title of first series of experiments.*

    A. Introduction.
    B. Materials and methods.
    C. Experiments and results.
    D. Discussion of results.

IV. *Descriptive title of second series of experiments.*

    A. Introduction.
    B. Materials and methods.
    C. Experiments and results.
    D. Discussion of results.

V. *Descriptive title of third series of experiments.*

    A. Introduction.
    B. Materials and methods.
    C. Experiments and results.
    D. Discussion of results.

VI. *General discussion.*

PROCESS OF WRITING

1. Mechanical Process. There probably is no best way to prepare a scientific paper, except as may be determined by the individual writer and the circumstances. With no notes at all, one might be able to start writing an article which, short or long, would be practically finished at every stage. Or one might accumulate the facts in a great mass of verbiage, and then, through condensation and revision, put the paper into final form. It is the end product that counts, not the intermediate steps.

2. Preliminary Outline. Most people obtain best results by developing a preliminary outline before they start writing. The following steps may be employed:

(*a*) Prepare a brief outline of the main topics to be treated in your article. Arrange the topics in a convenient and logical order. This outline may be on a single sheet of paper.

(*b*) Make a second, enlarged outline showing an analysis, by headings and subheadings, of the article. This may be three or four times as long as the first.

(*c*) Prepare a third outline before beginning the actual writing. In this outline the topics should be shifted to the most effective order, and each topic should be enlarged and preferably expressed in the form of a concise topic sentence.

(*d*) Begin the actual writing. Spread out before you the outline, the tables, and the graphs. Expand each topic or topic sentence of the outline into a paragraph. Make a rough draft first. Leave plenty of space between the lines for subsequent revision. Concentrate on the subject matter, and write as rapidly as possible, without letting details of language interrupt the flow of ideas. Then, on another day, examine critically what you have written and begin to revise it. (See section on "Revision," page 48.)

SUBJECT MATTER AND ARRANGEMENT

GENERAL

1. Unity. A scientific paper should be a unit, treating a single definite subject. It may contain several main topics if these are logical divisions of one large subject. Make a careful selection of materials. Include only what is necessary to an understanding of the main ideas, but omit nothing that is essential.

Each paragraph should have unity. It may well begin with a topic sentence that indicates the idea to be developed in the paragraph. The use of topic sentences aids the

writer in transforming his preliminary outline into paragraphs, and it helps the reader who looks through the
paper for its salient contents.

2. *Arrangement of Topics.* Choose a logical sequence of
topics, based upon a careful analysis of the subject matter.
The order may be determined by relations of space, time,
importance, similarity or contrast, complexity, or cause
and effect. Use an order that serves best the needs of clearness, coherence, and emphasis. Discuss similar points in
the same order, and use similar forms of expression. Indicate clearly the beginning of each new topic.

3. *Development.* Develop the main ideas until they are
clear enough to be easily understood by the reader. For
the sake of brevity in publication, it is usually necessary to
address the paper to specialists in the particular field,
rather than to the general reader. But a skillful writer,
while primarily addressing these specialists, can make his
paper intelligible and useful to many workers outside the
immediate narrow field.

Present the material in a manner that will enable the
reader to grasp its meaning as quickly and easily as possible. Use the style of the textbook—not that of the laboratory notebook. Explain each topic clearly, point by point.
Define, illustrate, prove, and summarize your statements,
if necessary.

Give considerable thought to the relative importance of
the various topics and their need of development. Treat
briefly those topics that are too simple to require detailed
explanation. Develop fully the more complex and the
more important topics. Achieve completeness and clarity
without sacrificing conciseness.

4. *Examples.* Illustrate the meaning of general or abstract statements by giving examples, particular instances,
concrete data, amplifying details, or specific comparisons.

5. *Reader's Questions.* Consider what questions the reader will wish answered in your article. Always keep in mind the fact that the primary purpose of your paper should be to give the reader valuable information and new ideas.

6. *Topics of General Interest.* Develop fully the topics that are of interest to many readers.

7. *Words.* Employ words that are approved by good usage. Be careful to avoid those that are obscure, ambiguous, or inappropriate. Prefer the specific word, the familiar word, the short word. Define all technical terms that the reader might not understand.

When at a loss for the word or expression that most precisely fits your thought, consult *Soule's Dictionary of English Synonyms, Roget's Thesaurus,* or *Webster's Dictionary of Synonyms.* And when in doubt regarding the idiomatic use of common words, turn to *The Concise Oxford Dictionary.* With a large vocabulary at your command, you can write concisely, because you can choose one word that will take the place of several in making your meaning clear.

Try to use words that a foreigner will be able to find in a small dictionary. If necessary, add a brief explanation of a word that might not be understood. (For example, a foreigner might not be able to find the meaning of the word "tumbler," but he would understand if you described it as a "cylindrical glass vessel, 7 cm in diameter and 10 cm deep.")

8. *Tone.* Skill may be developed in presenting material in a tactful way. Clear statements supported by evidence are better than positive assertions. Avoid pedantic or pompous language. Be careful also not to announce a well-known fact as if it were a discovery. Indicate clearly which of your results and conclusions are new. For completeness of discussion, it is often necessary to mention to the reader many things that he already knows; but this may be done skillfully, without annoying or confusing him.

TITLE

1. Choice of Title. Choose a concise descriptive title, complete enough to include the main topics needed for making a subject index in an abstract journal. Select these topics with the aim of giving definite ideas as to the exact contents of your paper. In a biological study, it is desirable to give the name of the organism in the title. Include all important nouns under which your paper should be indexed, but preferably limit the title to ten words. Place the more important words near the beginning of the title, so that the subject may be seen at a glance.

2. Selection of Topics. Ask yourself, "Under what topics would I look in a subject index of an abstract journal if I were searching for the literature on the subjects treated in my paper?" The answer to this question will provide the topics for your title.

INTRODUCTION

1. Content. The function of the introduction is to make clear the subject of the article. The introduction should state the problem, describe its condition at the beginning of the study, and tell the reasons for investigating it. It should give the purpose, scope, and general method of the investigation.

Finally, the introduction should state clearly and definitely the most significant result of the investigation. With the main conclusion before him at the start, the reader is able, as he goes through the paper, to judge the development of evidence and inference brought forward in its support. If, on the other hand, the statement of the main point is deferred until late in the paper, the reader is unable to distinguish essential from nonessential evidence and may overlook or forget important features.

2. Pertinent Literature. Make reference in the introduction to only those literature citations that bear directly upon the introduction itself. The other references to the literature should be included in the parts of the paper to which they are most pertinent, chiefly the discussion of results.

The foregoing procedure is now favored by most writers. To be sure, a long historical review—often arranged merely chronologically—was at one time considered to be an essential part of the introduction. But the reader generally finds such a review dull, since he is not prepared so early in the paper to correlate past investigations with the specific problem in hand. The place for most of the references to the literature is in the discussion of results, where the new results and interpretations are compared with those of previous investigators.

DISCUSSION OF RESULTS

1. Interpretation. The primary purpose of the discussion of results is to show the relations among the facts that you have observed. Support your conclusions by reference to the experimental results. Keep asking yourself, "What do the data show? What is the evidence?" (See the section on "First Steps in Treating Data," page 25.) Indicate the meaning of the facts, their underlying causes, their effects, and their theoretical implications. Aim, where possible, to explain facts in the symbols or language of mathematics, and according to the laws of physics and chemistry.

2. Reference to Tables and Graphs. Keep the text free from mere repetition of the detailed data presented in the tables and graphs. Such repetition, except when necessary to show comparisons not obvious in the tables and graphs, confuses the text and makes dull reading. As far as possible, the text should be reserved for comparisons, relations, conclusions, and generalizations.

3. Unsettled Points. Give particular attention to evidence that bears on points concerning which there is difference of opinion among scientists. But avoid personal or controversial language, or expressions likely to excite controversy or retort. Above all, do not impugn the motives of others; motives are irrelevant.

4. Emphasis of Conclusions. Indicate the ways in which the results of your study are related to the field as a whole. Emphasize the additions that it makes, and stress conclusions that modify in a significant way any hypothesis, theory, or principle that has secured general acceptance. Develop with special clearness observations or inferences that seem to be of sufficient importance to deserve mention in a textbook on the subject.

5. Qualification of Conclusions. To prevent misunderstanding, it is necessary to define as clearly as possible the precise conditions to which your conclusions apply. A conclusion should always be stated in such a way as to indicate its range of validity.

Confusion often results from failure to define adequately all influential experimental details. In any experiment or series of experiments the influential conditions may be analyzed conveniently into two groups: (1) those representing the variables specially studied, and (2) those representing the rest of the experimental complex—the influential background or prevailing conditions. The conditions of the first group are assumed to be adequately known and controlled. They are the conditions that are purposely made to differ in certain known ways. For an ideal experiment or experiment series, the conditions of the second group should be as thoroughly known and definitely described as are the primary variables. They should be maintained constant or at least not permitted to vary sufficiently to interfere with the influence of the primary variables.

6. Applications. Indicate the practical applications of your study to agriculture, industry, engineering, medicine, etc.

7. Stimulation. Try to stimulate the reader to further thought and research on the subject of your investigation.

ABSTRACT OR SUMMARY

1. Position and Designation. Two practices are followed in the various journals: (*a*) One is to print an *abstract* (often in distinctive type and without heading) at the beginning of the article, just below the title, where it is most convenient for readers. (*b*) The other is to print a *summary* (under this heading) at the end of the article.

This abridgment should be the same in content, whether it is designated as an abstract or as a summary.

Unfortunately, some journals print the abstract in type that is extremely difficult to read, because of small size, long lines, and close line spacing.

2. Purpose. In preparing a title and abstract for an article, it is important to realize that the individual worker glances over many more articles than he has time to read. A title is necessarily short but should be as informative as possible. In cases where the worker is uncertain from the title alone whether the article contains material of interest to him, the abstract helps him by telling more precisely what the article covers. Also in cases where he is interested only in the main results and conclusions, the abstract gives him this information in brief form.

The abstract fills a gap between the title, which may average only about ten words, and the article, which may be ten pages long. It is useful to readers who wish more information than is given by the title and less information than is given by the article. Its purpose, then, is to assist readers (a) by elaborating the title and (b) by condensing the article, thus saving the time of readers who do not require the full contents of the paper. Incidentally, if the abstract is well prepared by the author, it will be suitable for reprinting in an abstract journal.

3. Nature. To serve its purpose, the abstract should indicate clearly all the subjects dealt with in the article, so that no reader interested in only one of these subjects will fail to have his attention directed to it. The abstract should also summarize briefly but clearly the principal new results and conclusions, especially all new information likely to be of interest to readers who are not specialists in the field. The abstract should be well written, so as to be easily read and understood, and should be self-explanatory, complete and clear in itself.

4. Preparation. Keeping in view the dual purpose of the abstract, the writer should read his manuscript carefully, making notes (a) as to the subjects dealt with, particularly subjects concerning which new information is given incidentally, and (b) as to the new results and conclusions reported. Material relating to each subject should then be gathered together; sentences summarizing the material should be written; and finally these sentences should be put together so as to make a well-written abstract—brief, condensed, complete, yet readable.

5. Models. It will be useful to study as models the abstracts given in abstract journals and to try to make abstracts that would be acceptable to such journals.

44   WRITING THE PAPER

LOGICAL PRESENTATION OF IDEAS

Many of the mistakes in scientific papers involve errors
in logic. Their avoidance depends chiefly upon a thorough
understanding and careful analysis of the ideas that are
presented.[6] The following rules apply to some of the most
obvious, and yet commonest, mistakes of this type.

1. Requisites of a Good Hypothesis. A hypothesis is a
tentative explanation of certain observed facts. It is pro-
visionally adopted to explain these facts and to serve as a
guide for further investigation. The requisites of a good
hypothesis are the folowing: (a) It should explain facts
that have not hitherto been adequately explained. (b) It
should be consistent with all the known facts. (c) It should
be no more complex than necessary to account for the
phenomena. (d) It should aid the prediction of new facts
and relations. (e) It should be susceptible of verification or
refutation.

2. Illusions. (a) Be careful not to draw conclusions from
data involving errors of observation, errors in arithmetic,
compensating errors, systematic and personal errors. (b)
Do not use mathematical formulas without clearly understanding their derivation and all the assumptions involved. (c) Be cautious in comparing conclusions based
upon experiments in which the influential conditions
have been improperly controlled, and therefore not duplicated. (d) Avoid confusing facts with opinions or inferences, not only in the investigation itself but also in
preparing results for publication.

[6] "However skeptical one may be of the attainment of universal
truths, one can never deny that philosophic study means the habit
of always seeing an alternative, of not taking the usual for granted,
of making conventionalities fluid again, of imagining foreign states
of mind. In a word, it means the possession of mental perspective."—
William James.

3. *Too Broad Generalization*. (*a*) Do not draw a conclusion from too few data, nor too broad a conclusion from a limited series of data. (*b*) Be careful in drawing conclusions that are based on extrapolated curves. (*c*) Guard against failing to qualify a conclusion, so as to show the limits within which it applies, or the variation which is to be expected. (*d*) When you indulge in speculation, be sure that you, and your readers, know that it is speculation.

4. *Cause and Effect*. (*a*) Do not infer merely because one thing has followed another that it is the effect of the other. (*b*) Do not argue that causes are the same because indistinguishable effects have been observed. A certain phenomenon may have one cause in one case and another cause in a second case. (*c*) Be careful in making inferences by analogy. If two cases are observed to resemble each other in certain particulars, it is not safe to infer resemblance in another particular that has been noted in only one of them. (*d*) If two processes have the same mathematical expression (or yield the same sort of graph when plotted), it does not necessarily follow that the processes themselves are essentially alike.

5. *Prejudice*. (*a*) An attitude of intellectual honesty and devotion to truth is the foundation of scientific work. (*b*) Guard against prejudice; do not be influenced by preconceived opinions. (*c*) Do not decline to admit evidence because it necessitates an unwelcome conclusion. If a conclusion is unwelcome, it is a sign of a wrong mental attitude. (*d*) Biting, caustic comments are almost sure to be regretted later, and they invariably weaken the effect of one's arguments.

6. *Ambiguity of Terms*. (*a*) Guard against misunderstandings of language. (*b*) Define terms as clearly and precisely as possible. Do not use technical terms, especially in a field not strictly your own, unless you are certain of their precise meaning, or unless their use has been checked by a specialist in the field. (*c*) Do not use a term in one sense in one part of your reasoning and in another sense in another part. (*d*) Do not mistake a general for a specific use of a term. (*e*) Be very critical of statements containing the words *cause, determine, control, influence, result, effect*. Distinguish carefully between such words as *factor, condition, force, agency, process*.

7. Missing the Point. (*a*) Do not ignore the question, evade the issue, or argue beside the point. Define clearly the points at issue. Try to determine the crucial point that will really decide the discussion. (*b*) Guard against reasoning that may correctly prove something but not the thing which you think it proves.

8. Begging the Question. (*a*) Do not base a conclusion on an unproved proposition. (*b*) Avoid arguing in a circle—drawing a conclusion that merely states the assumptions in other words. (*c*) Do not assume the truth of a proposition that is not proved and may be false. (*d*) Do not assume that a certain thing is true because a prominent authority has said it is true. (*e*) Do not assume that a proposition is untrue because you are able to disprove the arguments that have been used to support it; there may be other, valid arguments that make it true.

MAKING THE PAPER INTERESTING

A mastery of the devices for attracting and holding the interest of the reader must be acquired by the writer of articles of a popular nature. These methods should be used cautiously by the writer whose purpose it is to inform, rather than entertain, his fellow-workers. Rather let your style be characterized by unobtrusive simplicity than by inappropriate and labored ornamentation. Content is more important than style. The writer should be more interested in the thing he is describing than in the words with which he describes it. Nevertheless, judicious use of some of the devices of the journalist may serve, without breach of propriety, to give a scientific or technical paper an attractive and interesting style. These devices are indicated by the following rules.

1. Begin with an introduction that is broad enough to give the reader the information necessary for an understanding and appreciation of the subject. Refer to the ways in which the subject is related to the reader's previous knowledge or experience, and suggest benefits to be derived from further information on the subject. Emphasize the economic or practical importance of the subject.

2. Make the paper as easy as possible for the reader to comprehend.

3. Use photographs, drawings, charts, diagrams, and curves.

4. Emphasize the new and the unusual—the features that have "news value."

5. Omit trivial and tedious details that are not essential for accuracy and completeness. Keep the text free from repetition of data presented in tables and graphs.

6. Link each part of the paper to some preceding part by transitional words, phrases, or sentences, so as to make a continuous story—thus sustaining the reader's interest.

7. Precede every dull passage by a stimulating introduction.

8. Seek variety in length and structure of words, sentences, and paragraphs.

9. Limit average sentence length to 20 words, and keep long words—those of three or more syllables—from exceeding 20 per cent of the total.

10. Use colorful words and vigorous turns of.expression.

11. Use forcible comparisons or resemblances. In popular articles and lectures occasionally use analogies, similies, and metaphors.

12. Introduce striking or unexpected statements, contrasts, and paradoxes.

13. Ask provocative questions.

14. Let the reader feel that he is doing his own thinking, not merely following. Stimulate his imagination and give him a sense of achievement.

REVISING THE MANUSCRIPT

After writing the first draft of your paper, begin to revise it. Revise several times, keeping one principal object in mind each time. Learn to rewrite between the lines, and to make corrections, insertions, and transpositions according to the methods given on pages 57 to 61. If there is not enough space between the lines for a revision, a convenient method is to write the revised passage on a slip of paper of page width and to staple this to the margin of the manuscript page. The pages need not be copied until they have become crowded with corrections.

Most writers will obtain best results by carrying a manuscript through one handwritten and two *triple-spaced* typewritten drafts, which allow plenty of space for revisions between the lines, before making the final double-spaced typewritten copy that is submitted for publication.

The writer is judged solely by the quality of his finished product. He should spare no effort in the revisions needed to make his manuscript as nearly perfect as possible.

1. Organization and Consistency. In the first revision, give attention to the order and development of the larger divisions of the paper—the sections and paragraphs. The order of the topics may need to be shifted, although this should not be necessary if a well-prepared analytical outline has been folowed. If the paper is long, the first part may have to be rewritten to make it consistent with the last part. Irrelevant parts should be eliminated. Important parts may be expanded, and minor parts subordinated.

2. Sentences. In the next revision of the rough draft of the manuscript, focus attention on the sentences. Many of these may need to be revised because they may have been written hurriedly, without much concern about details of form. Study and revise the sentences in groups, rather than singly. Make each group of sentences develop the exact ideas you wish to express. See that the members of the group stand in logical relation to one another. Achieve good organization of sentences through careful revision.

The following brief rules suggest helpful procedures:

(*a*) Use short sentences, with an average length of about 20 words, and a maximum length not exceeding 40 words (4 typewritten lines).

(*b*) Choose sentence structures that require only simple punctuation.

(*c*) Prefer to arrange the sentence elements in the normal order: subject, verb, object.

(*d*) Prefer the active voice of verbs.

(*e*) Keep the same subject and the same voice and use parallel structure.

(*f*) When advantageous, convert a loose compound sentence (consisting of statements joined by *and* or *but*) into a sentence with a subordinate clause.

(*g*) Transpose misplaced words, phrases, or clauses.

(*h*) Insert connectives and other reference words to show relations.

(*i*) Correct weak or vague reference of pronouns to their antecedents.

(*j*) Correct dangling modifiers (participles, gerund phrases, absolute phrases, prepositional phrases, infinitive phrases, and elliptical clauses).

3. *Clearness*. Revise sentences and paragraphs with special attention to clearness. There should be only one possible meaning, and this should be easily understood by the reader. Find a good word or phrase to convey your idea. A useful test of clarity is to see whether the paper is understandable when it is read rapidly aloud.

4. *Conciseness*. As a rule, the first draft of a paper should be longer and more complete than the copy that will be offered for publication. Better results are usually obtained by condensing a long paper than by expanding a short one. In shortening a paper, condense or eliminate the parts that are least needed for clearness of presentation. Strike out idle words (especially superfluous adjectives and adverbs); replace a phrase with a word; combine related sentences; eliminate repetition of an idea. Omit the obvious and the least important. Retain the essentials. Impartial counsel is valuable in aiding you to decide what is essential. In judging, put yourself in the place of the reader. It takes moral strength to "blue pencil" choice phrases, sentences, or paragraphs. But the results will justify the effort.

5. *Repetition*. Eliminate unnecessary or awkward repetition of a word or phrase. Some common "repeaters" in scientific articles are *found, showed, reported, studied, concluded, make, use, case, it is, there is, this is.* Scan the page, check the offenders with a red pencil, and then remove them. Substitute pronouns or synonyms, or preferably rewrite the sentences. Also avoid a succession of sentences of the same structure or length.

6. Connectives. Give special attention to connectives: *and, or, similarly, but, however, nevertheless, therefore, when, where, since, because, although, if,* etc.

7. Euphony. Revise to make the article pleasing in sound when it is read aloud. Eliminate a succession of the same sounds, and avoid words that rhyme. Reading the paper aloud rather slowly is a good way to catch various types of errors that need to be corrected.

8. Punctuation. Correct the punctuation. (See page 68.)

9. Style. Revise with special attention to typographical style, making certain that this conforms to that of the journal in which the paper is to be published. Consistency in the use of capitals and italics and in the style of headings is essential in a manuscript prepared for the printer's use. The printer cannot be expected to depart from the rule to "follow the copy."

10. Accuracy. Read through the manuscript carefully, searching for inaccuracy or exaggeration of statement.

11. Length of Paper. A paper may need to be shortened or divided to meet the limit specified by the journal in which it is to be published. (See section on "Estimating the Length of the Printed Article," page 62.) A long paper may often be divided into two or more short papers, and these may be published separately. Care should be taken, however, to make each paper a unit, treating one central topic. If there are two or more topics in a paper, these must be logical subdivisions of the single large topic.

12. Check List of Some Common Errors in Writing.

I. *Inaccuracy.*
1. Misstatement or exaggeration of fact.
2. Misrepresentation through omission of facts.
3. Errors in data, terms, citations.
4. Conclusions based on faulty or insufficient evidence.
5. Unreliable mathematical treatment.
6. Failure to distinguish between fact and opinion.
7. Contradictions and inconsistencies.

II. *Inadequate presentation.*

1. Omission of important topics.
2. Faulty order of sections or of paragraphs.
3. Inclusion of material in wrong section or paragraph.
4. Incomplete development of a topic.
5. Failure to begin a section or a paragraph with topic sentence containing the key word or phrase that indicates the subject.
6. Weak beginning of a section or a paragraph.
7. Inclusion of irrelevant, trivial, or tedious details.
8. Passages that are dull or hard to read.
9. Failure to distinguish between the new and well-known.
10. Inadequate emphasis of interpretation and conclusions.

III. *Faulty diction and style.*

1. Long sentences (more than 3 or 4 typewritten lines) and complicated grammar, or short, choppy sentences (less than 1½ typewritten lines).
2. Weak sentence beginnings—a string of weak or meaningless words.
3. Lack of clearness—a sentence that requires rereading to get the meaning.
4. Long, complicated paragraphs (more than ⅔ page of typewriting) or short, scrappy paragraphs (less than 5 lines of typewriting).
5. Wordiness and padding—failure to come directly to the point.
6. General words rather than definite words.
7. Dull, weak, or awkward expressions.
8. Unnecessary repetition of the same word or the same sentence structure.
9. Omission of connectives and other relation words, especially in short sentences.
10. Unnecessarily technical language or too many strange words in a single sentence.

PREPARING THE TYPEWRITTEN COPY

1. Copy for Typist. Copy for the typist should be clearly written in black ink or soft black pencil. Its style should conform in all details to that of the journal in which the paper is to be published. All the sheets should be of the same size, and numbered in the upper right-hand corner.

2. *One side of Paper*. Write on only one side of the paper.

3. *Flat*. Never roll a manuscript. Keep it flat, or, if necessary, fold it.

4. *Paper Clamps*. Fasten the sheets together with large paper clamps, which can easily be removed.

5. *Typewritten Manuscript*. The manuscript should be typewritten with a machine having clean type and a fresh, well-inked black ribbon. One should not risk having the paper rejected because the typing is difficult to read—small, faint, or blurred. A typewriter with pica characters (10 to the inch) makes much more legible original and carbon copies than one with elite characters (12 to the inch). The type should be kept clean; the carbon paper should be renewed frequently (after using for not more than about eight pages); and the ribbon should be changed as soon as the original copy becomes perceptibly lighter than the copy made with fresh carbon paper.

Triple-spacing is advantageous in the preliminary drafts of the manuscript because it allows plenty of room for revisions.

Double-spacing should be used throughout the final draft of the manuscript, including footnotes, legends, quotations, and literature citations. Single-spacing is permissible only where necessary to make a table fit the page.

White bond paper of standard size (8½ by 11 inches), good quality, and 16-pound weight should be used for the original copy.

6. *Number of Copies*. Three typewritten copies of the final draft of the manuscript should usually be made.

The author should retain one fully corrected carbon copy. The original copy (from the ribbon) should always be sent to the publisher, since a carbon copy is easily erased and may become illegible. Many journals require, in addition to the original typewritten copy, one or more carbon copies for examination by reviewers. For the convenience of the reviewers, these should be easy to read and handle. They may be made on slightly thinner paper than that used for the original—i.e., on 13-pound paper—but never on translucent, onionskin paper.

7. Margins. Leave a blank space of 2 inches above the title on the first page, 1 inch at the top of the other pages, and 1 inch at the bottom of each page. Leave a blank margin of 1¼ inches at the left side of each page and about 1 inch at the right side, but avoid dividing words at the ends of lines. Indent paragraphs 5 spaces.

8. Page Numbers. The pages of typewritten copy should be numbered consecutively in the upper margin, preferably in the right-hand corner.

9. Models of Style. The author should make a careful study of the journal in which his article is to be published, and he should prepare his copy so that it conforms to the best practice illustrated by recent issues of the journal. Only carefully prepared, clearly typewritten manuscripts are acceptable.

10. Directions for Proofs. The author's name and the address to which proofs are to be sent should be typewritten near the top of the first page of the manuscript and enclosed in a circle.

11. Title of the Paper. The full title of the paper, including the author's name, should be typewritten 2 inches from the top of the first page of the manuscript. The following example shows a complete heading that may be modified to suit the style of almost any journal. (The author's complete mail address should be printed in the paper, so that readers will know where to write for reprints. It is most convenient if given on the first page of the paper, in the heading or in a footnote to the title.)

[*Example of general heading*]

INFLUENCE OF SULFONAMIDES ON GROWTH AND RESPIRATION IN BACTERIA[1]

Henry E. Miller and John C. Stewart

Department of Bacteriology, School of Medicine, University of Pennsylvania, Philadelphia 4, Pennsylvania

Received for publication August 12, 1956

[1] The authors are indebted to Dr. Edward M. Johnson for helpful suggestions during the course of the study.

12. Footnotes, Citations, Quotations, Headings, Tables, Legends. See special directions for typewriting footnotes (page 91), citations (page 92), quotations (page 106), headings (page 111), tables (page 118), and legends (page 167).

It is essential that the manuscript be prepared in a way that will allow economical composition on a typesetting machine, which cannot compose two sizes of type in one operation. To permit rapid work, the manuscript should be arranged so that material to be printed in small type is on separate sheets that may be easily removed from the sheets that bear the text.

When the place for a table or for a quotation exceeding five lines is reached in typewriting a manuscript, the text sheet should be removed from the typewriter and a new sheet should be inserted. Only the table or quotation should be typewritten on this sheet. The text should be continued on a fresh sheet of paper.

Footnotes should not be typewritten with the text, but should be put on separate sheets (as many footnotes as convenient being written on a sheet). These sheets should be placed at the end of the text copy, after the literature cited. (Exception is made when preparing a manuscript for submission to journals that require footnotes to be placed in the text. See section on "Footnotes," page 91.)

The literature citations should also appear on a separate sheet.

The legends, or titles, of plates and figures should be typewritten in numerical order on one or more sheets, and these should be placed after the footnotes.

13. Title for Running Headlines. A condensed title of 35 letters or less should be given by the author for the running headlines of the pages. This should be placed on a separate sheet at the end of the manuscript.

CORRECTING THE TYPEWRITTEN COPY

1. Checking. After the manuscript has been typed, the author should read the typewritten copy for errors. All tables, figures, names, quotations, and citations in the copy must be verified by comparison with the original manuscript. A convenient method of checking is to have another person slowly read aloud from the original while you follow and correct the typewritten copy.

Assume that errors are present; find and correct them. The responsibility for uncorrected errors in figures, names, citations, and quotations rests entirely with the author, since the publisher has no means of discovering such errors. It is fatal to leave them for critics to discover after the paper has been published.

2. Indicating Special Characters. The typewritten manuscript must be clear to the typesetter, who is not a scientist. Symbols, signs, superscript letters and figures, etc., must be unmistakable. The symbol "Cl" (for chlorine) must be marked with a handwritten "ℓ" above it to show that it is not "Cl" if the typewriter uses the same symbol for both the letter "l" and the figure "1". This should be done even if the two symbols are almost identical in some type faces. The multiplication sign "×" must be plainly marked or the words "multiplication sign" written in the margin, to distinguish it from the letter "X". All but the simplest mathematical expressions must be clearly printed by hand. Greek letters or other unusual characters should be clearly written and, if necessary, explained by marginal notes. Write "OK as typed" above a word that might be regarded as mistyped. An ordinary dash (em dash) should be typewritten as two hyphens, without space before, between, or after them. If a hyphen occurring at the end of a typewritten line should be printed as a hyphen, mark it "=".

CORRECTIONS

1. Corrections in Body of Manuscript. If possible, write corrections in the body of the manuscript, not in the margin. If corrections are written in the margin, it may be difficult to make necessary transpositions by cutting and pasting. Do not destroy legibility by writing too many words between the lines. When it is necessary to reconstruct a long sentence or a paragraph, typewrite the revision upon a separate slip of paper of page width, and paste this directly over the section rewritten. (Staples are more convenient than paste, but they should never be used in the draft sent to the printer.)

2. Corrections Horizontal. Write corrections horizontally on the page.

3. *Corrections Above Line.* Place the corrections in the space above the line to which they apply so that the printer will see them before he reaches the words concerned.

4. *Cancellation.* To cancel a word or a sentence, draw a horizontal line through it. To cancel a single letter, draw a vertical line through it.

5. *Restoration.* To restore a word that has been canceled by mistake, rewrite the word above the one you have canceled, or make a series of dots under the word and write "Stet" in the margin.

6. *Substitution.* To replace one word by another, cancel the first word by drawing a horizontal line through it, and write the new word immediately above. Never write the new word directly upon the first.

7. *Indicating a Paragraph.* When a word should begin a new paragraph, place the "¶" sign immediately before the word.

8. *Canceling a Paragraph.* To cancel a paragraph division, write "No ¶" in the margin, and draw a "run-in" line from the indented word to the last of the preceding sentence.

9. *Period.* A period may be indicated clearly by enclosing it in a small circle.

10. *Space Between Words.* To separate two words that have been written together, draw a thin vertical line between them.

11. *Canceling Space Between Words.* To indicate that two words are to be brought together, connect them by means of half-circles above and below them. (For example: Foot ͜ note.)

12. *Reduction of Capital Letter.* To indicate that a capital letter should be printed as a small (lower-case) letter, draw through it a light oblique line sloping downward from right to left.

13. Italic Capitals. Four lines under a letter or word indicate printing in *ITALIC CAPITAL* type.

14. Capitals. Three lines under a letter or word indicate printing in ROMAN CAPITAL type.

15. Small Capitals. Two lines under a letter or word indicate printing in SMALL CAPITAL type.

16. Italics. One straight line under a letter or word indicates printing in *italic* type.

17. Boldface. One wavy line under a letter or word indicates printing in **boldfaced** type.

INSERTIONS

1. Brief Inserts. To insert one word or a few words, write them above the line and indicate the place for their insertion by a caret (∧) placed below the line.

2. Longer Inserts. To insert a passage of several or many lines in a page of the manuscript, this method of cutting and pasting is recommended: Write the passage on a fresh piece of paper of page width. Cut the manuscript page at the place where the insertion is to be made; arrange the material in the proper sequence, with the surplus going on a new page; and paste the pieces of paper on full-sized sheets. Give the added manuscript page an interpolated number. For example, if it follows page 10, give it the number "10A"; at the bottom of page 10 write "Follow with page 10A," and enclose this note in a circle. If several additional pages are required, number these "10A," "10B," "10C," etc.

Another method that may be used is the following: Write the passage on a separate slip of paper of page width. Mark this slip "Insert A," and draw a circle around the passage. In the margin of the manuscript page write "Insert A," draw a circle around it, and from the circle draw a line to a caret at the place where the insertion is to be made. Paste the slip to the margin of the page. Several inserts in the same page are marked "Insert A," "Insert B," "Insert C," etc. If full-sized sheets are used, instead of slips of paper, give them interpolated page numbers, following the manuscript page.

TRANSPOSITIONS

1. Transposition by Cancellation and Insertion. The simplest way to transpose one word or a few words is to cancel them, and then write them in their new position above the line, with the place for their insertion indicated by a caret (∧) placed below the line.

2. Transposition on the Same Page. To indicate transposition of a passage to a different place on the same page, draw a circle around the passage, and from the circle draw a line to the margin and continue the line to a caret at the new place.

3. Transposition from One Page to Another. The best way to transpose a passage to another page is to cut the manuscript, arrange the material in the desired sequence, and paste the pieces of paper on full-sized sheets. Then, if necessary, renumber the pages.

An alternative method is the following: Suppose the passage is to be transposed from page 15 to page 14. Draw a circle around the passage on page 15, and write in the margin "tr A to p. 14." In the margin of page 14 write "tr A from p. 15," draw a circle around this note, and from the circle draw a line to a caret at the desired place for the passage. Other transpositions to page 14 are designated "B," "C," etc.

RENUMBERING PAGES

1. Consecutive Page Numbers. The methods given above refer to insertions and transpositions made in the preliminary drafts of an article.

Before the manuscript is submitted to an editor, or sent to a printer, all the material must be in proper sequence on full-sized sheets that are numbered consecutively.

2. Cutting and Pasting. Any insertions or transpositions that are necessary should be made by cutting, rearranging, and pasting on full-sized sheets. It is not necessary to have all the sheets filled with typewriting. The pages should be renumbered by canceling the original numbers and writing the new numbers near the canceled ones. There should be no interpolated page numbers (such as 7A, etc.).

FINAL REVISIONS

1. Finished Manuscript. The author is expected to make all final revisions in the typewritten manuscript. Corrections cost nothing if they are made in the manuscript, but alterations in the proofs are very expensive and are likely to introduce inconsistencies and new errors.

2. Permissible Corrections. A manuscript in which there are no corrections often indicates a careless author. If the changes are not too many and are made clearly, it will not be necessary to retype the pages.

3. Order of Material. Before sending your manuscript to a publisher, be sure to have all parts in the proper order, as outlined below:

Author's name and address to which proofs are to be sent.

Title, name of author, footnote to title.

Text material (each table and each long quotation on a separate page).

Literature cited (on a separate page).

Footnotes (on a separate page).

Legends for illustrations (on a separate page).

Condensed title of 35 letters or less (on a separate page).

Copy for illustrations.

ESTIMATING THE LENGTH OF THE PRINTED ARTICLE

1. Formula and Method. An accurate estimate of the length of the text of the printed article can be made by means of the following simple formula:

Number of printed pages =
$$\frac{\text{Characters per MS line} \times \text{Lines per MS page} \times \text{Pages of MS}}{\text{Characters per printed line} \times \text{Lines per printed page}}$$

Letters, punctuation marks, and spaces between words are counted as characters. Short lines at ends of paragraphs are counted as full lines.

For example, suppose that the manuscript has an average of 63 characters per line, an average of 27 lines per page, and a length of 23 pages; and that the printed page has an average of 52 characters per line and has 124 lines per page.

$$\text{Number of printed pages} = \frac{63 \times 27 \times 23}{52 \times 124} = 6.1$$

Allowance must be made for the space to be occupied by tables and illustrations. This may be difficult to estimate accurately. If center headings are numerous, they must be taken into account also.

The space required for a legend may be calculated in a similar manner, by taking into account the number of characters per line of such material and the number of lines per vertical inch on the printed page. (See also page 125.)

This method of estimating the length of printed material is easier and much more accurate than any method based upon word count. Words vary in length from "a" or "if" to "nitrobenzenesulfonamides." So the number of words per line is much more variable than the number of characters. Character count is the basis of the system used by printers for copy-fitting.

2. *Character Count in Typewriting.* The average number of characters (including blank spaces) per manuscript line is obtained as follows: (*a*) Measure the length in inches of the average line on each of ten pages; (*b*) obtain the mean of these measurements; and (*c*) multiply the mean by 10 for pica typewriting or by 12 for elite typewriting. (If the typewriting has unequal spacing of the characters, multiply by the average number per inch.)

The average number of lines to the page is determined by counts made on representative pages. Double-spaced typewriting has 3 lines to the inch, for both pica and elite.

Accurate estimation of typewritten material is facilitated by a ruler (Seneca Secretary), obtainable at stationery stores, which has scales that measure directly the number of pica or elite characters per line and the number of lines per page.

3. *Character Count in Printing.* Obtain the average number of characters per printed line as follows: (*a*) Count the number of characters (letters, punctuation marks, and spaces between words) in each of ten full lines picked at random in the journal in which your article is to be published; and (*b*) obtain the mean of these numbers. The number of printed lines per page is determined by a count made on a full page.

KINDS OF TYPE

1. *Roman.* The light-faced, vertical type in general use is called roman. There are three kinds of roman type: (*a*) CAPITALS (caps), which may be indicated in the manuscript by drawing three lines under the word or letter to be capitalized; (*b*) SMALL CAPS (capital letters about half as high as caps), which may be indicated in the manuscript by drawing two lines under the letter or word; (*c*) lower-case letters (ordinary small letters). A diagonal line may be drawn lightly through a capital letter to indicate that it should be printed as a lower-case letter.

2. Italics. *In italic type, or italics, the letters slope up toward the right.* To indicate italic type, draw a single straight line under the letter, word, or figure. If italic capitals are desired, underscore with four straight lines.

3. Boldface. Type with a conspicuous or heavy face is called **boldface** or **blackface**. To indicate this type, underscore with a wavy line. Its principle uses are for headings in textbooks and for names of new species of plants and animals.

4. Face and Body. A single piece of type cast by a Monotype machine is a rectangular block of metal with a flat top that bears, in relief, a letter or other character. The upper or printing surface of the raised character is the face, and the block bearing the character is the body. The part of the flat top that projects beyond the base of the raised character is known as the shoulder. A whole line is cast in one piece, or slug, by a Linotype machine.

5. Sizes of Type. The size of the type is designated by the height of the rectangular top of the smallest body on which the face can be cast. The following examples illustrate common sizes, as they appear when printed:

This line is set in 6-point type.

This line is set in 8-point type.

This line is set in 9-point type.

This line is set in 10-point type.

This line is set in 11-point type.

This line is set in 12-point type.

The unit employed in sizes of type is the point, or $\frac{1}{72}$ inch. Thus 10-point type has a body 10 points ($\frac{10}{72}$ inch) high, and has a face, or raised character, slightly less in height, so that there will be a very small space between the printed lines. When 10-point type is used in composition with only this small space between the lines, it is said to be set "solid." Usually, however, the lines are separated by the additional space provided by casting the type on a larger body. In most work 10-point type is cast on a 12-point body. The type is then said to be 10-point type on 12-point body. This book is printed in type of that size, with subsidiary matter in 8-point on 10-point and the Index in 8-point on 9-point.

Scientific journals that have only one column on a page usually employ 10-point type on 12-point body or 11-point on 13-point, with subsidiary matter (footnotes, legends, literature citations, tables, etc.) set in 8-point on 10-point.

Journals with a two-column format are usually printed in 9-point or 10-point type on 10-point or 11-point body, with subsidiary material in 8-point on 9-point.

6. Size of Type Page. The unit employed in measuring the width and height of the type page is termed a pica (also called em pica or 12-point em), which equals $\frac{12}{72}$ or $\frac{1}{6}$ inch. This unit was originally based upon the width of the capital letter "M" of pica or 12-point size. Thus the type page of this book is $3\frac{1}{2}$ inches or 21 picas in width ($3\frac{1}{2} \div \frac{1}{6} = 21$) and $5\frac{9}{16}$ inches or 34 picas in height (excluding running headlines).

7. Spacing. The em is used as a unit for measuring printed matter. In most standard type styles, an em of 12-point type is 12 points ($\frac{1}{6}$ inch) wide (and also 12 points high), an em of 10-point type is 10 points wide, and an em of 8-point type is 8 points wide.

The em and halves of the em are used for indentation

and spacing, and also for expressing the lengths of dashes. An em quad is a block of type that is one em in width; the ordinary dash (—), or em dash, is the width of an em quad. An en quad is half of the width of an em, and an en dash (–), used to separate page numbers in citations, is an en in width.

8. Specifications and Printing. Complete specifications for a publication include the styles and sizes of type for text, subsidiary matter, tables, citations, headings, etc., the dimensions (in picas) of the type page (both including and excluding running heads), the composition or type-setting, the proofreading, the handling of illustrations, the margins, the make-ready (preparation for printing), the paper, the press work for the number of copies required, the binding, the wrapping and mailing, etc. The publisher ordinarily takes care of these details, but an editor or business manager of a journal, or an author who is preparing copy directly for the printer, needs to give considerable attention to these matters. It is evident that many complicated factors are involved in the printing process. Close cooperation with the printer and proper preparation of the manuscript will effect significant savings in printing costs without sacrificing legibility or appearance.

Chapter 3

GOOD FORM AND USAGE

TENSES

1. Experimental Facts. The experimental facts should be given in the *past tense*. (For example: The plants *grew* better in A than in B. The dry weight *was* greater in A than in B.)

2. Presentation. The remarks about the presentation of data should be mainly in the *present tense*. (For example: Diagrams of dry yields *are* shown in figure 3. The second column of table 2 *represents* the dry weight of tops.)

3. Discussions of Results. Discussions of results may be in both the *present* and *past tenses,* swinging back and forth from the experimental facts to the presentation. (For example: The highest dry weight *is* shown for culture A, which *received* the greatest amount of the ammonium salt. This may mean that the amount of nitrogen added *was* the determining condition for these experiments.)

4. Specific Conclusions. Specific conclusions and deductions should be stated in the *past tense,* because this always emphasizes the special conditions of the particular experiments and avoids confusing special conclusions with general ones. (For example: Rice *grew* better, under the other conditions of these tests, when ammonium sulfate *was* added to the soil. Do not say: Rice *grows* better when ammonium sulfate *is* added to the soil.)

5. General Truths. When a general truth is mentioned, it should, of course, be stated in the *present tense*. Logically, a general truth is without time distinction. For example, one may say, "Many years ago, scientists were convinced that malaria *is* caused by a germ carried by a certain species of mosquito." Well-established principles of mathematics, physics, and chemistry should be put in the *present tense.*

PUNCTUATION[7]

Good punctuation is an aid in achieving clarity and emphasis in writing. Punctuation should follow current usage and should be uniform throughout an article. It is better to learn to apply a few simple rules than to puzzle over each case as a separate problem. Punctuation then becomes almost automatic. The occasional perplexing case is best treated by putting the sentence into a form that requires only simple punctuation to make its meaning immediately clear. The following general rules of punctuation are most frequently applied:

1. End Punctuation. Put a period, a semicolon, or a colon—usually a period—at the end of a complete declarative statement (containing subject and predicate). A period is never incorrect, but in some cases a semicolon or a colon helps to indicate relations between statements.

A semicolon may sometimes be used to advantage at the end of a short statement that is followed by a closely related statement of similar length.

Science and the arts in different ways pursue a common end; they are expressions of man's effort to bring order and beauty and understanding into his life. —*Graham DuShane.*

Insight and imagination are needed to formulate probable hypotheses; logical and sometimes mathematical power is needed to deduce their consequences; patience, perseverance, and experimental skill are needed to test their validity. —*W. C. Dampier-Whetham.*

Although semicolons provide a means of grouping logi-

[7] Also consult the books by Perrin (1950), Summey (1949), and Woolley (1909).

cally related statements, they give less conspicuous breaks than periods, and so retard reading. The trend in current writing is toward almost complete elimination of semicolons.

Use a period or a semicolon—preferably a period—after a statement followed by one introduced by a conjunctive adverb, such as *however, yet, still, nevertheless, so, therefore, consequently, accordingly, hence, moreover, also, thus, likewise, furthermore, then, indeed, otherwise.*

The solution of a scientific problem does not close a chapter; it opens new problems. Moreover advances in one field of science make possible advances in another. —*C. E. K. Mees.*

Since the longer conjunctive adverbs tend to give a rather formal style, most writers use them only sparingly, and usually prefer to place them, enclosed in commas, in the middle of sentences.

In this sense a law of nature expresses a mathematical relation between the phenomena under consideration. Every physical law, therefore, can be represented in the form of a mathematical equation. —*J. W. Mellor.*

A colon is preferable to a semicolon or a period at the end of a statement which formally points to one that follows.

2. Coordinate Statements. Always put a comma, a semicolon, or a period—usually a comma—before an independent statement (containing subject and predicate) introduced by one of the pure coordinating conjunctions—*and, but, for, or, nor.*

What we know about the world depends on what we can do, and this depends on the instruments at our disposal. —*Justus von Liebig.*[8]
The seeds of great discoveries are constantly floating around us, but they only take root in minds well prepared to receive them. —*Joseph Henry.*[8]

[8] Reprinted by permission from *Science Digest.*

If either of the statements joined by the conjunction is complex and contains commas within itself, a semicolon or a period should be used—a period if the statements are long. A period is required if it is desirable or logically necessary to have the statement introduced by the conjunction stand out distinctly from the preceding statement.

False facts are highly injurious to the progress of science, for they often endure long. But false views, if supported by some evidence, do little harm, for every one takes a salutary pleasure in proving their falseness. —*Charles Darwin*.[8]

Of all the servants of morality, science is the greatest; for it is the one serious way we have to discover what means are likeliest to lead to the realization of the ends we cherish. —*W. R. Dennes*.

The scientist should not be a mere technician; he must be a wise member of society. Nor can society be well guided by men who are ignorant of those criteria for reaching sound conclusions that are the essence of science. —*J. H. Hildebrand*.

3. Series of Coordinate Elements. A comma should precede *and* in a series of coordinate elements such as *a, b, and c,* in which the elements may be words or phrases.

The scientific method requires of its practitioners high standards of personal honesty, open-mindedness, focused vision, and love of the truth. —*Warren Weaver*.[8]

4. Adverbial Clauses. Always put a comma after an adverbial clause that precedes its principal clause. But separation by a comma is usually unnecessary when the adverbial clause follows the principal clause. Adverbial clauses are introduced by *when, while, before, after, until, where, because, since, if, unless, although, though,* etc.

Comma necessary

Until man has no more curiosity and no more wants, his quest for knowledge will persist and the application of that knowledge will continue. —*C. E. K. Mees*.

While we can coax physical nature into satisfying many of our wishes, we cannot exercise authority over it or make it change its ways one jot. —*Bertrand Russell.*[8]

Comma unnecessary

The brightest flashes in the world of thought are incomplete until they have been proved to have their counterparts in the world of fact. —*John Tyndall.*[8]

Nature will tell you a direct lie if she can. —*Charles Darwin.*

5. Relative Clauses. A nonrestrictive relative clause is not essential to the meaning of the sentence; it should always be set off by commas. A restrictive relative clause is essential to the meaning; it should never be set off by commas. Relative clauses are generally introduced by *that, which, who,* or *whose.* (Many writers prefer *that* to *which* for restrictive relative clauses.)

The test is to omit the clause in reading the statement. If it can be omitted without changing the meaning of the statement, it is nonrestrictive and should be set off by commas.

Nonrestrictive

Science is not the mere collection of facts, which are infinitely numerous and mostly uninteresting, but the attempt of the human mind to order these facts into satisfying patterns. —*C. N. Hinshelwood.*

Restrictive

There is no law of physics that cannot be overturned by an individual on his own initiative if he can prove a set of facts that contradict it. —*Vannevar Bush.*[8]

6. Erroneous Junction. Use a comma or a dash to separate two parts of a sentence that might be erroneously joined in reading. For example, always set off by a comma an introductory phrase that contains a verb in some form, such as an infinitive or a participle.

With accurate experiment and observation to work upon, imagination becomes the architect of physical theory. —*John Tyndall.*

Viewed historically, it would seem that respect for fact is more difficult for the human mind than the invention of theories. —*Bertrand Russell.*[8]

We must remember that we cannot get more out of the mathematical mill than we put into it, though we may get it in a form infinitely more useful for our purpose. —*J. Hopkinson.*

Science has given us the means either to exalt or degrade humanity—perhaps even to destroy civilization. —*H. F. Guggenheim.*[8]

7. *Interpolated Elements.* Enclose in commas, dashes, or parentheses an interpolated element that would make the meaning of the sentence obscure if no punctuation were used.

No scientist, however great his renown, can mislead his fellow scientists for longer than it takes to check his observations or verify his conclusions and their consequences. —*I. I. Rabi.*[8]

It is ideas that enable us, not only to find our way among the myriad facts of any one area, but even, now and then, to take excursions into neighboring territory. —*J. H. Hildebrand.*

Like the philosopher, the scientist is required, not so much to record the way things happen as time goes on, as to decide on the merits of different ways of thinking about these things. —*Stephen Toulmin.*[8]

CAPITALS

The subject of capitalization is difficult to handle with definite rules, but capitals should be used according to a uniform style throughout a single article. For this reason a special revision of the manuscript should be made with the aim of making capitalization uniform.

1. *Proper Nouns.* Capitalize a proper noun, designating an individual person or thing. Also capitalize a derivative of a proper noun if the derivative retains close association with the proper noun.

2. Words Derived from Proper Nouns. Be consistent in the capitalization of words derived from proper nouns. The words *volt, ampere, farad, ohm, coulomb,* and *watt* should not be capitalized. It is better to capitalize *India ink, Paris green, Prussian blue, plaster of Paris, Bordeaux mixture.* Follow consistently a single dictionary, preferably *Webster's New International Dictionary* or *Webster's New Collegiate Dictionary.*

3. Manufactured Products. Capitalize the significant parts of the name of a manufactured product. (For example: Pyrex glass, Cellophane membrane, Amberlite resin.)

4. First Words. Begin with a capital: a sentence, a complete sentence directly quoted, a legend of a table or an illustration, a center subheading, a paragraph side heading, or a topic in a table of contents.

5. Titles of Publications in Text. In the text, capitalize all important words in titles of books and periodicals and in titles of chapters in books and of articles in periodicals. (For example: Chapter XII of Clark's *The Determination of Hydrogen Ions* is entitled "Theory of the Hydrogen Electrode." An article on "Cobalt and Nickel in Soils and Plants" appeared in *Soil Science.*)

In footnote citations and in lists of literature cited, capitalize only the first word and proper nouns in English titles of books and of articles in periodicals (page 96).

6. Scientific Names. In botanical and zoological work, capitalize the scientific names of genera, families, orders, classes, subdivisions, and divisions of plants and animals. (For example: *Triticum,* Gramineae, Glumiflorae, Monocotyledoneae, Angiospermae, Spermatophyta.)

7. Common Names Derived from Scientific Names. Do not capitalize common names derived from scientific names of plants and animals. (For example: amoeba, angiosperm, bacillus.)

8. *Chemical and Medical Terms.* Do not capitalize the names of chemicals, medicines, diseases, and anatomical parts.

9. *Table, Figure, Plate.* Do not capitalize *table, figure,* and *plate,* unless this is the style of the journal for which you are writing. (For example: The results given in table 2 are shown as graphs in figure 3.)

10. *Miscellaneous Terms.* Do not capitalize such words as *plot, plat, series, class, exhibit, form, group, schedule, section, appendix, station,* etc., even when immediately followed by a figure or a capital letter, except when the style of the journal calls for capitals.

ITALICS

Indicate italic type in the manuscript by underlining with a single straight line the letters, words, or numerals that are to be italicized.

1. *Algebraic Symbols.* Letters used as algebraic symbols should be italicized. (For example: $Ax + By + C = 0$.) For economy in composition of type, some publications italicize only the full-sized letters, not superscript letters. Numerals should not be italicized. Subscript letters should usually not be italicized.

Letter symbols for units of measurement should not be italicized, even when used in mathematical expressions. (For example: $1 \text{ ev} = 1.601 \times 10^{-12} \text{ erg}$.)

2. *Explanatory Letters in Illustrations.* Some journals prefer to use italic or slant letters to designate points, lines, objects, etc., in diagrams, drawings, and graphs. Even if roman or vertical lettering is used in the illustration, italics should always be used in the legend and in the text when reference is made to such explanatory letters. (Example of legend of diagram: Fig. 1. Diagrammatic cross section of coconut pinna, lines AB and AC representing the two pinna wings, hinged to the midrib at A.)

3. Genera and Species. In the text of botanical, bacteriological, zoological, and geological work, italicize scientific names of genera, species, and varieties, and of genera alone. [For example: *Phaseolus lunatus*; *Musa sapientum* Linn. var. *cinerea* (Blanco) Teod.; *Escherichia coli*; *Phytophthora*.] But do not italicize names of classes, orders, and families.

When used in literature citations, indexes, tables, and titles, scientific names are usually not italicized. Many zoological publications do not italicize scientific names even when used in the text.

4. Common Names Derived from Scientific Names. Do not italicize common names derived from scientific names of plants and animals. (For example: amoeba, angiosperm, bacillus, bacterium, paramecium, protozoan, streptococci.)

5. Books and Periodicals. Italicize titles of books, pamphlets, and periodicals when these appear in the text. (For example: Fieser and Fieser's *Organic Chemistry*.) In footnote citations and in lists of literature cited, such titles are usually not italicized.

6. Subdivisions of Books and Periodicals. Use quotation marks, not italics, for titles of chapters in books or titles of articles in periodicals when these are given in the text. (For example: Chapter 1 of Yost and Russell's *Inorganic Chemistry* deals with "Nitrogen and Its Oxides and Sulfides." An article on "Absorption of Water by Plants" appeared in the *Botanical Review*.) In footnotes and in lists of citations, it is customary to use neither italics nor quotation marks.

7. Article. The word *the* or *a* should be italicized and capitalized when it begins the title of a book, but not when it begins the title of a periodical. (For example: Fisher's *The Design of Experiments*. An article in the *American Journal of Botany*.)

8. Technical Terms. It is permissible to italicize a letter or word to which special attention is called. An unusual technical term, requiring formal definition, may be italicized the first time it appears in an article. When an expression is regarded as quoted, it should be enclosed in quotation marks. (For example: The term *atmometric index* will be used in place of the expression "evaporating power of the air.")

It is best to avoid over-use of italics, capitals, and other special devices for emphasizing ideas. They often lead to an exaggeration of an idea or fact. If used excessively, they do not even give emphasis or distinction.

9. Chemical and Medical Terms. Do not italicize the names of chemicals, medicines, diseases, and anatomical parts. (For example: Uranium hexafluoride, hydroquinone, atropine, penicillin, diabetes mellitus, esophagus.)

10. Foreign Words. It usually is better not to italicize foreign words. (For example: Intra-vitam staining, ceteris paribus, in medias res, in situ, en masse, e.g., i.e., viz., et al.) Some journals, however, italicize foreign words that have not come into common use in English.

NUMBERS

1. General. Use figures for all *definite* weights, measurements, percentages, and degrees of temperature. (For example: 6.7 kg, 2¾ in., 15.6 ml, 112°C.) Spell out all *indefinite* and *approximate* periods of time and all other numerals that are used in a general manner. (For example: One hundred years ago, thirty years old, about two and one-half hours, ten instances, three times.) Judgment must be exercised in this matter. For instance, figures should be used in experimental data where periods of time are definite and of frequent occurrence. The conservative rule is to spell out numbers wherever possible. Some journals spell out only small numbers, those under 10 or under 100.

2. Consistency. Be consistent throughout the article in the use of figures. Do not express small numbers in words in one paragraph and in figures in another.

3. Beginning of Sentence. Never begin a sentence with a figure. Revise the sentence; or, if this is impossible, write the number in words.

4. Avoiding Confusion. Spell out numbers if confusion would be caused by the use of figures. (For example: Fifteen 200-watt Mazda lamps.)

5. References to Tables. Use figures in the text for all numbers taken from tabular matter.

6. Metric System. The metric system of weights and measures should usually be employed in scientific publications. Where it is customary to use a nonmetric system, as in engineering, it may be desirable to give metric equivalents in parentheses.

7. Abbreviations. Universally understood abbreviations of metric weights and measures may be used in tables and in footnotes, and in the text when directly following figures. Nonmetric units should usually be spelled out except in engineering. (For a list of abbreviations and the rules for using them, see page 80.)

8. Temperatures. Temperatures should be expressed in Celsius (centigrade) degrees except where it is customary to use Fahrenheit degrees.

9. Time. Employ figures for hours of the day, using a colon to separate hours and minutes. (For example: 7:00 a.m.; 3:30 p.m.; 12 m.; 12 p.m.)

10. *Dates.* Use figures for days of the month, spelling out the name of the month and omitting *d, th, st.* (For example: On September 21, 1956; in September, 1956. Put a comma after the year number unless it ends a sentence. A form without punctuation is: 21 September 1956.)

11. *Money.* Use figures for sums of money written with a dollar sign. (For example: $15.65; $25, *not* $25.00; but definite precision sometimes requires the use of ciphers at right of decimal.)

12. *Twenty-one to Ninety-nine.* Cardinal numbers from twenty-one to ninety-nine, inclusive, should be written with hyphens.

13. *Hyphens in Ordinal Numbers.* Ordinal numbers should be joined with hyphens. (For example: Thirty-fourth, one-hundred-and-eleventh.)

14. *Comma in Figures.* In tabular matter, use a comma to separate a number of four or more figures, grouping three units to the right. In the text, omit a comma in a number containing four figures.

15. *Large Numbers.* Large numbers may be indicated by a means of exponents. (For example: 4.39×10^6 instead of 4,390,000.)

16. *Fractions.* Decimal fractions should be employed in the metric system. Common fractions used in an indefinite manner should be spelled out, joining the numerator to the denominator with a hyphen. (For example: One-half of the balance, two-thirds of the residue, about one-tenth of this quantity.) Use figures for common fractions when designating definite weights and measurements. (For example: ½-in. pipe.) Large fractional expressions may be written with a slant line or solidus (29/32). Fractions that would require very large numbers in numerator or denominator should be expressed decimally. Very small fractions are conveniently indicated by means of negative exponents. (For example: 7.5×10^{-8}.)

17. Half and Quarter. Compounds of *half* and *quarter* should be written with a hyphen. (For example: Half-full; quarter-past. But: One half was dried; the other was not.)

18. Per Cent. Omission of a period after *per cent* is favored by most writers. (Some journals use the symbol % in tables or even in the text.)

19. Per cent and Percentage. Do not use *per cent* for *percentage. Per cent* should be preceded by a number. (For example: Three analyses gave the following percentages of sugar: 93.2, 93.1, and 92.9. There was an increase of 15 per cent in production.)

20. Basis for Percentage. If there is possibility of misunderstanding, make clear the basis used for expressing percentages. (For example: The phrase "a 5 per cent solution of alcohol in water" correctly means 5 grams of alcohol in 100 grams of the solution, but some writers use it to mean 5 ml of alcohol in 100 ml of the solution. Also, in reporting analyses of foods, plant and animal tissues, blood, milk, etc., it is essential to specify whether moisture-free weight, fresh weight, or volume is used as the basis for percentages.)

21. Standard Error or Probable Error. State whether *standard error* or *probable error* is meant when an expression such as "10.3 ± 0.2 gm" is used.

22. Plural. Use the plural form when referring to a quantity or measurement of more than one. (For example: About one and one-half kilometers; 1¼ inches.)

23. Singular and Plural Forms of Verbs. When total quantity is indicated, the singular verb may be used. (For example, it is permissible to write: To each culture 300 ml of solution was added.) But it is better to recast the sentence and avoid the difficulty. (For example: Each culture received 300 ml of solution.)

24. *Mathematical Expressions.* To simplify printing, reduce mathematical expressions to a single line when possible. Use a slant line or a negative exponent to signify division, and use fractional exponents instead of square-root and cube-root signs.

25. *Verification.* The use of statistical or mathematical formulas should be checked by a specialist in the field.

26. *Significant Figures.* In publishing a computed number, retain no more significant digits than are consistent with its accuracy. In statistical work the following rule is a useful guide: In the published constant, retain no figures beyond the position of the first significant figure in one-third the standard error; in all computations, keep two more places. (For example: 129 ± 3, *not* 129.2 ± 3.1.)

27. *Roman Numerals.* Where possible, avoid the use of Roman numerals, since they are not readily understood.

SYMBOLS (NAMES OR ABBREVIATIONS) FOR UNITS OF MEASUREMENT[9]

The symbols—names or their abbreviations—that are chosen for units of measurement should be suitable for use in mathematical equations in the same manner as numerical symbols and letter symbols for quantities. The rules for selecting and using unit-symbols should there-

[9] Reprinted, with minor changes, by permission of the author and publisher from an article in *Science*. (*Roller, D. 1954. Science 120: 1078–1080.*)

See also the American Standards Association's (1941) *Abbreviations for Scientific and Engineering Terms* and the special rules that govern the use of abbreviations in the journal in which your paper is to be published.

fore be formulated according to established mathematical procedures. Recommended symbols for many units of measurement are given in table 2.

In general, it is a good rule to use only those abbreviations which you know are used by careful writers in your field of science or technology, and to conform to the style of the publication in which your article is to be published. With a few exceptions, the forms recommended here are those widely used.

SELECTION OF SYMBOLS FOR UNITS

Rules are given below for the selection of symbols for units of measurement. These rules facilitate a choice between several commonly used symbols, and they provide a basis for the selection of a symbol for a new unit. Exceptions to the rules are made in the cases of some symbols that are firmly established by usage.

1. *The symbol should preferably consist of two or three letters, and never more than four.* Thus "lu" is preferable to "l" for "lumen." Well-established exceptions to this rule include "gamma" for "gamma," "j" for "joule," and "v" for "volt." If the name of a unit contains only two or three letters, this name, rather than an abbreviation, should be used; examples are "bar," "day," "erg," "lux," and "ohm".

2. *The symbol should preferably consist of the first two or three letters of the name of the unit.* This facilitates recognition and pronunciation, as in "dy" for "dyne," "lu" for "lumen," and "oer" for "oersted." Among the firmly established exceptions to this rule are "ct," "ft," "hr," "hy," "lb," and "oz".

3. *There should preferably be only one symbol for a particular unit.* But it seems desirable to recommend certain exceptions, as illustrated by the following ex-

amples: both "a" and "amp" for "ampere," and "b" and "bel" for "bel"—the "a" and "b" to be used only with combining forms, as in "ma" and "db".

As the rule implies, the same symbol should be used for both singular and plural forms; thus "10 amp" (not "10 amps").

4. *Periods should be omitted from symbols.* An exception is "in.," for "inch," since omission of the period would often result in confusion.

5. *The symbol for a combining form denoting a multiple or submultiple should be a single letter.* Thus "μ" for "micro-," "M" for "mega-," and so on.

USE IN PRINTED TEXT OF SYMBOLS FOR UNITS

1. *Set the symbol for a unit in roman type (not italic).*

2. *Use an abbreviated unit-symbol only when it:* (a) *is preceded by a numerical value, or* (b) *appears in headings of tables or in crowded text, in which cases the symbol is enclosed in parentheses.* Thus: "25 cm"; "several centimeters"; "volumes, in cubic centimeters"; "volumes (cm^3)"; "v (cm^3)."

3. *Identify or explain any symbol that might not be immediately clear to the readers of the paper.* In some cases it is advisable to do this in a glossary in the introduction to the article.

4. *Use standard signs to indicate all mathematical operations with unit symbols.* Thus indicate: (*a*) multiplication by a space, a center dot, or even $\times$; (*b*) division by a solidus, a negative exponent, or the ordinary fractional form; (*c*) a power by a positive exponent; (*d*) a root by a fractional exponent or $\sqrt{}$. Thus: "dy cm," "dy·cm," or even "dy $\times$ cm"; "ft/sec," "ft sec^{-1}," or "$\frac{ft}{sec}$"; "cm^3".

TABLE 2

Symbols for units of measurement

UNIT OR COMBINING FORM	SYMBOL	EXAMPLES OF USE AND COMMENTS
ab-	ab	abamp, abcoul/cm^2
acre	acre	acre ft
ampere	amp	amp turn/m
	a	With prefixes: ma, μa
angstrom	A	
are [$\equiv 10^2$ m^2]	are	Use sparingly
atom	atom	atom/gm-awu, atom/mole
atomic mass unit	amu	1 amu = 931 Mev
atomic weight unit	awu	1 awu = 1.0002 amu
atmosphere, standard	atm, A_s	
atmosphere at 45°	atm_{45}, A_{45}	
bar [$\equiv 10^6$ dy/cm^2]	bar	
barn [$\equiv 10^{-24}$ cm^2]	barn	
barrel	bbl	
barye [$\equiv$ dy/cm^2]	$\cdots$	Use μbar or dy/cm^2
bel	bel	
	b	With prefixes: db-
billion [$\approx 10^9$ or 10^{12}]	$\cdots$	Ambiguous; deprecated
Bohr magneton, electronic	μ_0	1 μ_0 = 9.27 $\times$ 10^{-21} erg/gauss
Bohr magneton, nuclear	μ_1	1 μ_1 = μ_0/1836 erg/gauss
Brinell hardness number	Bhn	
British thermal unit	Btu	Btu/lb F°
bushel	bu	bu/acre
calory	cal	
Calory	$\cdots$	*See* kilocalory
candle	ca	ca hr, ca/m^2
candlepower	$\cdots$	*See* candle
Celsius degree (*temp. difference*)	C°	cal/cm sec C° 100C° $\approx$ 180F°
Celsius temperature		*See* degree Celsius
cent (*acous.*)	cent	
cent (*monetary*)	ct	ct/gal, ct/kw hr
	¢	In crowded tables, etc.

TABLE 2.—*Continued*

UNIT OR COMBINING FORM	SYMBOL	EXAMPLES OF USE AND COMMENTS
centi- $[\equiv 10^{-2}]$	c	cm
centigrade	· · ·	*See* Celsius
centimeter-gram- second unit	cgsu	Use sparingly
centimeter-of- mercury	cm-hg	
centimeter-of-oil	cm-oil	
cgs electromagnetic unit	cgsm	*See also* ab-
cgs electrostatic unit	cgse	*See also* stat-
circular (*adj.*)	cir	cir-mil $[\equiv 0.7854 \text{ mil}^2]$
coulomb	coul	coul/m^2
count	count	count/min
cubic	³	cm^3 (never "cc"), ft^3/sec
curie	c	mc, μc
cycle	cy	cy/sec $[\equiv$hz$]$
	c	Often used with prefixes: kc, Mc
day	day	
deca- $[\equiv 10]$	· · ·	Deprecated
deci- $[\equiv 10^{-1}]$	d	db, dm
degree (*of arc*)	deg	deg/sec (*ang. velocity*)
	°	90°
degree absolute	· · ·	*See*: degree Kelvin; degree Rankine
degree Baumé	°B	
degree Celsius	°C	0°C$\equiv$32°F
degree Fahrenheit	°F	32.000°F (*ice point*)
degree Kelvin	°K	273.16°K
degree Rankine	°R	491.69°R
deka- $[\equiv 10]$	· · ·	Deprecated
diopter	diop	
division	div	div/sec, div/μv
dollar	dol	dol/hr, dol/ton
	$	In tables, etc.
dozen	doz	doz/hr
dyne	dy	dy/cm^2, dy cm (*torque*)
electromagnetic unit	emu	Ambiguous: deprecated
electron (*charge*)	e	

TABLE 2.—*Continued*

UNIT OR COMBINING FORM	SYMBOL	EXAMPLES OF USE AND COMMENTS
electron volt	ev	1 ev = 1.601 × 10⁻¹² erg
electrostatic unit	esu	Ambiguous; deprecated
erg	erg	erg sec, erg/C°, erg/deg
Fahrenheit degree (*temp. difference*)	F°	Btu/F°
farad	fd	fd/m
	f	With prefixes: abf; μf
foot	ft	ft/sec
foot-candle [≡lu/ft²]	ft-ca	lu/ft² is preferable
foot-lambert [≡ca/π ft²]	ft-lam	
foot-pound-second unit	fpsu	Use sparingly
fresnel [≡10¹² cy/sec]	fr	
gallon	gal	gal/min
gamma [10⁻⁵ oer]	gamma	
gamma [≡μg]	. . .	Deprecated; use microgram, μg
gauss	gauss	
geepound	. . .	Synonym for slug
giga- [≡10⁹]	G	Gev
gilbert	gil	
gill	gill	
grad [≡10⁻² rt. angle]	grad	
grain	gr	
	grain	In medical work
gram	gm	
	g	With prefixes: kg; mg
	gram	In medical work
gram atomic weight	gm-awu	
gram calory	. . .	*See* calory
gram molecular weight	. . .	*See* mole
grav [≡32.174 ft/sec²]	grav	
Hartree unit	hu	
hecto-, hect- [≡10²]	h	Use sparingly: hm
hekto-	. . .	Variant of hecto-
henry	hy	hy/m
	h	With prefixes: mh

TABLE 2.—*Continued*

UNIT OR COMBINING FORM	SYMBOL	EXAMPLES OF USE AND COMMENTS
hertz [$\equiv$cy/sec]	hz	hz sec [$\equiv$cy]
horsepower	hp	hp hr
hour	hr	hr/day
	h	Astron. text and tables: 3^h
inch	in.	in./sec
inch-of-mercury	in.-hg	
inch-of-oil	in.-oil	
joule	j	j/mole deg
Kelvin degree (*temp. difference*)	K°	
	deg	Use when not ambiguous: erg/deg molecule
kilo- [$\equiv 10^3$]	k	kcal, kev, kmole, kg
kilogram-calory	$\cdots$	*See* kilocalory
kilocalory	kcal	
kilogram-mole	$\cdots$	*See* kilomole
kilomega- [$\equiv 10^9$]	kM	*Also see* giga-
kilomole	kmole	
knot	knot, ķn	knot hr, kn hr
lambert [$\equiv$ca/π cm^2]	lam	
large calory	$\cdots$	*See* kilocalory
light-year	lt-yr	
line	line	line/cm^2
liter	lit	
	l	With prefixes: ml
lumen	lu	lu hr, lu/watt
lux [$\equiv$lu/m^2]	lux	lux sec
magnetic pole	$\cdots$	*See* pole
magneton	$\cdots$	*See* Bohr magneton
maxwell	max	
mega-, meg- [$\equiv 10^6$]	M	Mev, Mm
megamega- [$\equiv 10^{12}$]	$\cdots$	*See* tera-
megohm [$\equiv 10^6$ ohm]	meg, Mohm	meg/m, Mohm/m
meter	m	
meter-candle [$\equiv$lux]	m-ca	lux is preferable
meter-kilogram-second unit	mksu	Use sparingly
mho	mho	mho/cm

TABLE 2.—*Continued*

UNIT OR COMBINING FORM	SYMBOL	EXAMPLES OF USE AND COMMENTS
micro-, micr- [$\equiv 10^{-6}$]	μ	μsec, μv, μg
micrometer	$\cdots$	*See* micron
micromicro- [$\equiv 10^{-12}$]	$\cdots$	*See* pico-
micron [$\equiv 10^{-6}$ m]	μm, μ	
mil [$\equiv 10^{-3}$ in.]	mil	
mile	mi	mi/gal
mil-foot	mil-ft	ohm/mil-ft
milli- [$\equiv 10^{-3}$]	m	mm, ma, ml (not "cc")
millimicro-[$\equiv 10^{-9}$]	$\cdots$	*See* nano-
millimicron [$\equiv 10^{-9}$ m]	mμ	
million [$\equiv 10^{6}$]	M	Mgal/day; *see* mega-
minute (*of arc*)	min	min/sec (*ang. velocity*)
	$'$	In tables, etc.
minute (*of time*)	min	
	^{m}	Astron. text and tables: 5^{m}
mks electromagnetic unit	mksm	
mole	mole	
molecule	molecule	
month	mo	
myria- [$\equiv 10^{4}$]	myria	Use sparingly
nano- [$\equiv 10^{-9}$]	n	
neper [$\equiv 8.686$ db]	nep	
newton [$\equiv$ kg m/sec^2]	new	new/m^2
normal atmosphere	$\cdots$	*See* atmosphere, standard
number	no.	no./hr
ohm	ohm	ohm cm
	Ω	In crowded diagrams, etc.
oersted	oer	
ounce	oz	
phon	phon	
phot [$\equiv$lu/cm^2]	phot, ph	
pico- ($\equiv 10^{-12}$]	p	
pint	pt	
pole, unit magnetic	pole	dy/pole
poise [$\equiv$dy sec/cm^2]	poise	
pound	lb	

TABLE 2.—*Continued*

UNIT OR COMBINING FORM	SYMBOL	EXAMPLES OF USE AND COMMENTS
poundal	pdl	ft pdl
pulse	pulse	pulse/sec
quart	qt	
radian	rad	rad/sec
Rankine degree (*temp. difference*)	R°	Btu/lb R°
revolution	rev	rev/min, rev/sec ("rpm" and "rps" only in crowded tables, etc.)
rhe [≡1/poise]	rhe	
rod	rod	
roentgen	r	mr
rowland	row	
r-unit	...	Synonym for roentgen
rutherford	rd	
second (*of arc*)	sec	
	″	In tables, etc.
second (*of time*)	sec	
	ˢ	Astron. text and tables: 10ˢ
slug [≡32.174 lbm]	slug	slug/ft³ (*density*)
Siegbahn unit	...	Synonym for x-unit
small calory	...	*See* calory
square	²	in.²
stat-	stat	statamp, statcoul
steradian	srad	
stilb [≡ca/cm²]	...	Deprecated; use ca/cm²
tera- [≡10¹²]	T	
thousand [≡10³]	k	kBtu, kft
ton	ton	
turn	turn	
vibration	vib	vib/sec
volt	v, volt	v coul, volt coul
watt	watt	watt hr
	w	With prefixes: kw
weber	web	web/m²
x-unit	xu	1 xu = 1.0020 mA
week	wk	
yard	yd	
year	yr	

THE GREEK ALPHABET

A α	Alpha	I ι	Iota	P ρ	Rho
B β	Beta	K κ	Kappa	Σ σ s	Sigma
Γ γ	Gamma	Λ λ	Lambda	T τ	Tau
Δ δ	Delta,	M μ	Mu	Υ υ	Upsilon
E ϵ	Epsilon	N ν	Nu	Φ ϕ	Phi
Z ζ	Zeta	Ξ ξ	Xi	X χ	Chi
H η	Eta	O o	Omicron	Ψ ψ	Psi
Θ θ ϑ	Theta	Π π	Pi	Ω ω	Omega

NAMES OF PLANTS AND ANIMALS

PLANTS

1. Complete Name. A complete plant name should include the name of the genus (in italics), the name of the species (in italics), and the abbreviated designation of the person who named the plant (in roman type). (For example: *Oryza sativa* Linn.) It is often desirable to add the common name of the plant, and in some cases the name of the family (both in roman type). [For example: *Shorea polysperma* Merr. (tanguile), Dipterocarpaceae; *Hemileia vastatrix* Berk. & Br. (coffee rust), Pucciniaceae. If two authors are involved, some authorities prefer the ampersand or "et" to "and."]

Unfortunately, a plant may have received several common and scientific names. Where scientific names differ in standard or commonly used works, one is chosen and the others are treated as synonyms. If a synonym is much used, it is customary to insert it in parentheses after the accepted name. In an index, accepted names are usually printed in roman type and synonyms in italics. In tables and in titles, names of genera and species are usually printed in roman type.

2. Necessity of Scientific Name. The scientific name, in addition to the common name, should be given when the plant is first mentioned in a paper. Use names that will be understood by foreign readers. For example, *Manihot utilissima* is universally understood by botanists, but the common name camoteng cahoy would be unintelligible to readers in most parts of the world. The scientific name may be enclosed in parentheses after the common name. [For example: The experiments described in this paper deal with the growth of rice (*Oryza sativa* Linn.).]

3. Use of Common Name. In papers dealing with agriculture, the scientific name of a well-known plant need not be repeated. After the scientific name has been given once, the plant may be referred to by its common name in the rest of the paper.

4. Capitalization. The generic name should be capitalized, but the specific name should usually not be capitalized. There is good authority, however, for capitalizing names of species derived from generic names, or from names of persons. (For example: *Acer Negundo, Ustilago Zeae, Magnolia Soulangeana.*)

5. Variety Name. Capitalize the vernacular names of plant varieties (Yellow Dent corn, Binocol rice, Carabao mango, New Era cowpeas), but not the latinized names of varieties (*Lathyrus palustris* Linn. var. *linearifolius* Ser.).

ANIMALS

1. Complete Name. In papers on zoology or one of its branches, such as entomology, names of animals should usually be given in a form similar to that used for plant names. [For example: *Agromyza destructor* Malloch (bean fly), Family Agromyzidae, Order Diptera; *Bubalus bubalis* Lyd. (carabao), Bovidae; *Equus caballus* Linn. (horse), Equidae.] Many zoological publications, however, do not italicize scientific names.

2. *Use of Common Name.* Well-known kinds of animals may be referred to by their common names. The complete scientific name may be given only at the beginning of the paper, or it may be omitted entirely. (For example: Berkshire swine, Hereford cattle, horse, rabbit, rat, Barred Plymouth Rock fowls.)

FOOTNOTES

1. *Reference Numbers in Text.* Footnotes pertaining to the text should be numbered consecutively (from 1 up) throughout each article and indicated by superscript numerals ([1, 2, 3,] etc.). The reference numeral to the footnote should be placed in the text after the word or sentence to which the footnote refers. It is placed *after* a punctuation mark if one occurs. Indicate the superscript numeral by typewriting it above the line and placing a V-shaped mark under it. Observe that these references apply to the text only; tabulations employ a separate series of symbols or superscript letters for each table. If mathematical formulas containing exponents appear in the text, care should be taken to avoid confusing exponents and footnote reference numbers.

2. *Footnotes at End of Manuscript.* Footnotes should not be in the body of the text; the text should have the reference numbers only. Footnotes should be typewritten *double-spaced* on one or more separate sheets (as many footnotes to a sheet as convenient). Each footnote should be indented as a paragraph, and preceded by a superscript numeral corresponding to the reference number in the body of the manuscript. The sheets bearing footnotes should be put at the end of the text copy, each sheet bearing the word "Footnotes," enclosed in a circle.

This method is necessary in order to facilitate composition on the typesetting machines. When printed, each footnote will be inserted at the foot of the proper page.

Exception is made when a manuscript is being prepared for submission to a journal which requires that the footnotes be placed in the text of the manuscript. In this case each footnote is inserted (double-spaced) in the text, beginning a new line immediately following the line of text containing the reference numeral, and the footnote is set off by rules from the text material above and below it.

3. Misuse of Footnotes. Use footnotes only where they are indispensable. They are expensive and distract attention from the text. A sentence in parentheses may often take the place of a footnote. Include important material in the text; omit irrelevant material.

LITERATURE CITATIONS

Citations to the literature are given in a list at the end of the paper or in footnotes distributed through the paper. The method of handling text references to citations and of printing the citations differs in detail in the various journals. Editors and publishers may eventually agree upon a uniform standard for citations. The standard form should be one that can be easily handwritten and typewritten, and requires no editorial marking for the printer. It should therefore avoid the use of small capitals, italics, and blackfaced type—all of which make extra work for author, typist, editor, and printer. Until such a standard is adopted, however, the author must make his citations conform to the particular style of the journal in which his paper is to be published.

1. Making a Card File. In writing citations in the library, follow the exact style employed by your journal, and take time to check every item. Include the full title (with accents and other diacritical marks) for your own use, even if you will eliminate it in the final draft of your manuscript. Put only one citation on a card. Small cards (3 by 5 inches) are more convenient to handle, but large cards (5 by 8 inches) have the advantage of providing space for an abstract of the article.

It is best to keep the citations in the form of a card file until the rest of the manuscript is ready for the final typing. This permits ready insertion or elimination of citations during the revisions of the manuscript, and is especially advantageous if the citations are to be numbered serially throughout the article. Finally, the citations are typewritten from the cards on sheets of manuscript paper.

2. Verifying the Citations. Verify each item in every citation by going to the library or the reprint file and looking up all the publications. Many errors result from failure to check citations taken from literature lists. As each citation is checked, make a clear notation on the card, so that doubt will not arise later. You must assume full responsibility for the accuracy and completeness of your citations. Although the editor may make minor revisions in the form of the citations to suit the style of his journal, he cannot be expected to correct spelling, figures, etc., or supply missing data.

3. The Heading. When the citations are printed at the end of the paper, the heading "Literature Cited" or "References" is usually employed. Only citations that are specifically referred to in the text are included in such a list. It is customary to use the heading "Bibliography" in books or articles of a general or popular nature in which specific reference to all the citations is not made in the text.

4. Directions for Two Methods. Directions are given below for two methods of handling citations. Each is widely used, with minor modifications, in scientific and technical literature. The first method has the following advantages not possessed by the second: (*a*) Reference by author and year of publication gives the reader the information he wants in the text and enables him to locate the citation easily in the alphabetical list at the end of the paper, or to use the list independently as a source of literature. (*b*) This method allows citations to be inserted or removed, during the revisions of the manuscript, without the necessity of repeatedly renumbering the series. (*c*) Since the citations contain no small capitals, italics, or blackfaced type, they are easily handwritten and typewritten, and require no editorial marking.

An editor or publisher may adopt for his journal the first method, the second method, or some modification of either. A contributor, however, must follow the particular style used by the journal to which he submits the paper.

FIRST METHOD

1. Text Reference to Citation. Reference in the text to a citation is made by means of the author's name followed by the year of publication in parentheses. [For example: Foster (1954).] If the paper cited has more than two authors, reference may be made by adding "et al." to the name of the first. [For example: Smith et al. (1956).] Where the author's name does not form a part of a sentence in the text, reference is made in parentheses after the proper word or at the end of the sentence. [For example: (Bailey, 1952), (Allen, 1950; Dodge, 1938; Thompson, 1956).] If reference is made to several papers published in the same year by one author, the suffixes a, b, c, etc., are used after the year number, the suffixes being chosen according to the order of reference in the text.[10]

---

[10] Where greater brevity is required, reference to citations is made

2. Arrangement of Citations. The citations are type-written, double-spaced throughout, from the card file. They begin on a new sheet of paper, at the end of the article, bearing the center heading "Literature Cited," in capitals.

The citations are arranged alphabetically according to authors' names. The author's name is typewritten flush with the left-hand edge of the writing, and second and succeeding lines are indented 5 spaces on the typewriter. A number of papers by the same author are listed in chronological order, according to the year of publication; several papers in one year are given the suffixes a, b, c, etc., after the year number. Some journals use a long dash in place of repetition of the author's name. In case of multiple authorship, the name of the first author usually determines the alphabetical and chronological order in the list.[10]

Journals with numbered volumes

Each citation of a paper in a journal includes the following items:

(a) *Surname of author* followed by a comma and initials. (For example: Williams, R. R.) If there are several authors, only the name of the first is inverted, for alphabetizing. (For example: Garner, W. W., and H. A. Allard.) Note that a comma is required after the inverted initials.

---

in the text by numerals in parentheses. The numeral is placed after the author's name, or after the proper word or at the end of the sentence. To shorten the text, the names of many of the authors are omitted. The numerals in the text refer to citations at the end of the article, which are numbered in the order of text reference, or preferably alphabetized and then numbered.

In the early drafts of the manuscript, it is advantageous to use the author's name and the year of publication as the reference in the text. Just before the final typing, the cards are numbered serially and the text references are changed to numerals, as described above.

(*b*) *Year of publication* followed by a period. (For example: 1956.) If several papers published in the same year by one author are cited, the year number is followed by a, b, c, etc., in the order of reference in the text. (For example: 1956a, 1956b, 1956c.)

(*c*) *Title of paper,* exactly like the original in wording and punctuation. A period follows the title. Only proper names are capitalized, except in Danish, Dutch, or German. Accents and other diacritical marks are added when used in foreign titles. (If extreme brevity is required in the citation, the title of the paper is omitted.)

(*d*) *Abbreviated name of serial publication* in the approved form. (See section on "Abbreviations of Periodical Publications," page 100.)

(*e*) *Volume number* followed by a colon.

(*f*) *Page numbers.* The number of the first page of the paper is separated by an en dash (indicated by a hyphen) from the number of the last page, and the latter is followed by a period.

Bucher, W. H.   1950.   The crust of the earth. Scient. Amer. 182 (5): 32–41. (The issue number is given in parentheses after the volume number if each issue is paged separately.)

Hartford, C. G., M. R. Smith, and W. B. Wood, Jr.   1946.   Sulfonamide chemotherapy of combined infection with influenza virus and bacteria. Jour. Exper. Med. 83: 505–518.

Liverman, J. L.   1955.   The physiology of flowering. Ann. Rev. Plant Physiol. 6: 177–210.

Morginson, W. J.   1946.   Toxic reactions accompanying penicillin therapy. Amer. Med. Assoc., Jour. 132: 915–919.

Rhoades, M. M.   1938.   Effect of the Dt gene on the mutability of the a_1 allele in maize. Genetics 23: 377–397.

Taylor, D. L.   1942.   Influence of oxygen tension on respiration, fermentation, and growth in wheat and rice. Amer. Jour. Bot. 29: 721–738.

Books

The copyright date (usually on the reverse of the title page) is used as the year of publication. The following examples illustrate the form used:

Baldwin, E. 1952. Dynamic aspects of biochemistry. 2nd ed., 544 p. Cambridge, England: Cambridge University Press.

Bayliss, W. M. 1931. Principles of general physiology. 4th ed., 882 p. London: Longmans, Green and Co.

Breed, R. S., E. G. D. Murray, and A. P. Hitchens. 1957. Bergey's manual of determinative bacteriology. 7th ed. Baltimore: Williams and Wilkins Co.

Friedlander, G. 1955. Nuclear and radiochemistry. 468 p. New York: John Wiley and Sons.

McElroy, W. D., and B. Glass. (Ed.) 1955. A symposium on amino acid metabolism. 1048 p. Baltimore: Johns Hopkins Press.

Yearbooks

An example illustrates the form used:

Matthews, A. F. 1948. Uranium and thorium. Minerals Yearbook 1946: 1205–1231.

Yearbooks are not numbered as volumes, but only by years. The actual time of publication—as shown in the example—is usually one or two years later than the period covered by the yearbook.

Experiment station bulletins

In citing experiment station bulletins and other issues of serial publications bearing an individual number but no volume number, the following form is used:

Beath, O. A., H. F. Eppson, and C. S. Gilbert. 1935. Selenium and other toxic minerals in soils and vegetation. Wyoming Agric. Exper. Sta. Bull. 206: 1–55.

Hutchinson, G. E. 1950. Survey of contemporary knowledge of

biogeochemistry. 3. The biogeochemistry of vertebrate excre-
tion. Amer. Mus. Nat. Hist. Bull. 96: 1–554.

SECOND METHOD

1. Footnote Citations. Footnote citations are numbered consecutively in the paper and are referred to in the text by superscript numerals. If other footnotes occur (except those in tables), they are numbered in the same series with the citations. A repeated reference is given the number of the original reference. The superscript reference numeral is placed in the text after the word or sentence to which the footnote refers. It is put after a punctuation mark if there is one. The superscript numeral is indicated by typewriting it above the line and putting a V-shaped mark in pencil or ink under the numeral.

Each footnote citation is inserted as a separate line (or lines) immediately following the line of text containing the word to which it refers. The footnote is indented as a paragraph and preceded by a full-sized numeral in parentheses (or in some journals by a superscript numeral without parentheses). It is typewritten double-spaced and set off by short rules from the text material above and below it.

For economical composition on typesetting machines, many journals require that the footnotes be typewritten on separate sheets (as many as convenient on a sheet). The sheets bearing the footnotes are put at the end of the text copy, each marked with the word "Footnotes," enclosed in a circle.

2. Citations at End of Article. In some journals full-sized numerals in parentheses in the text refer to citations at the end of the article. The citations are numbered in the order of the text references, or are alphabetized and then numbered.

Journals with numbered volumes

Each citation of a paper in a journal includes the following items:

(a) *Reference number*—full-sized numeral in parentheses or superscript numeral, according to the style of the journal.

(b) *Initials and surname of author* followed by a comma.

(c) *Abbreviated name of serial publication* in italic type, indicated by typewritten underlining with a single straight line, followed by a comma.

See section on "Abbreviations of Periodical Publications," page 100. Note also that many chemical journals use the abbreviations given by *Chemical Abstracts* in its "List of Periodicals Abstracted," and that most medical journals follow *Index Medicus*. Abbreviations from *Chemical Abstracts* are used in the examples given below.

(d) *Volume number* followed by a comma, both in blackfaced type, indicated by underlining in pencil or ink with a wavy line.

(e) *Number of the first page of the article.*

(f) *Year of publication of the article* in parentheses followed by a period.

(1) P. E. Porter, C. H. Deal, and F. H. Stross, *J. Am. Chem. Soc.*, **78**, 2999 (1956).

(2) E. C. Stoner, *Phil. Mag.*, [7] **36**, 803 (1945).

(3) E. H. Rhoderick, *Proc. Roy. Soc. (London)*, **A201**, 348 (1950).

(4) (a) R. W. G. Wyckoff and R. B. Corey, *Z. Krist.*, **89**, 462 (1934); (b) R. D. Waldron and R. M. Badger, *J. Chem. Phys.*, **18**, 566 (1950).

Books

In citing books, the form shown by the following examples is used:

(1) J. W. Mellor, "A Comprehensive Treatise on Inorganic and Theoretical Chemistry," Longmans, Green and Co., New York, N.Y., 1931, vol. 11, p. 341.

(2) L. F. Fieser and M. Fieser, "Organic Chemistry," third edition, D. C. Heath and Co., Boston, Mass., 1956, p. 475.

(3) R. D. Evans, "The Atomic Nucleus," McGraw-Hill Book Co., New York, N.Y., 1955, p. 802.

(4) W. Heitler, "The Quantum Theory of Radiation," Clarendon Press, Oxford, England, 1954, pp. 216–230.

Yearbooks

The following example illustrates the form used:

(1) N. B. Melcher, *Minerals Yearbook*, 1946, 751 (1948).

Experiment station bulletins

In citing experiment station bulletins and other issues of serial publications bearing an individual number but no volume number, the form shown by the following example is used:

(1) H. B. Vickery and G. W. Pucher, *Conn. Agr. Exp. Sta. Bull.*, 352 (1933).

ABBREVIATIONS OF PERIODICAL PUBLICATIONS[11]

In preparing literature citations, the author should base abbreviations of serials upon a careful study of those used by the publication in which his paper is to be printed. Capitalization, in particular, varies in different journals. A uniform style must be used throughout a single bibliography.

The rules given below are intended to make it as easy as possible for the readers of your article to look up the literature references in the library.

1. Beginning with Key Word in Library Card Catalogue. The abbreviated name of a periodical publication

[11] Miss Margaret C. Shields, of the Fine Hall Library of Princeton University, helped in the preparation of this section.

should always begin with the key word under which the name is entered alphabetically in the library card catalogue and other library lists.

The sequence of words in the abbreviated name should therefore follow the rules used in the *Union List of Serials in Libraries of the United States and Canada*. This reference volume is available in most libraries. With permission of the publisher, the H. W. Wilson Co., these rules are quoted below:

(*a*) "A serial not published by a society or a public office is entered under the first word, not an article [*a, an, the,* or equivalent], of the latest form of the title."

Annual Review of Biochemistry	Ann. Rev. Biochem.
Annalen der Physik	Ann. d. Physik
The Botanical Review	Bot. Rev.
The Journal of Experimental Biology	Jour. Exper. Biol.

(*b*) "A serial published by a society, but having a distinctive title, is entered under the title, with reference from the name of the society."

American Journal of Botany (published by the Botanical Society of America)	Amer. Jour. Bot.
Chemical and Engineering News (published by the American Chemical Society)	Chem. and Engineer. News
Science (published by the American Association for the Advancement of Science).	Science

(*c*) "The journals, transactions, proceedings, etc., of a society are entered under the first word, not an article, of the latest form of the name of the society."

This rule applies to publications of a *society*, an *association*, an *academy*, an *institution*, or a *university*.

If the name of the society (or other organization) is put at the beginning of the abbreviated title of such a publication, the reader will naturally look under that name in the library card catalogue, and so will easily find the publication. But if the word "Bull.," "Jour.," "Proc.," or "Trans.," is placed at the beginning of the abbreviated title, the reader is likely to waste much time searching in the library catalogue under that word.

If all authors and editors would adopt this library rule and adhere to it consistently, they would save their readers much loss of time.

Academie des Sciences, Paris.
    Comptes Rendus.................... Acad. des Sci. Paris, Compt. Rend.
American Chemical Society.
    Journal........................... Amer. Chem. Soc., Jour.
American Medical Association.
    Journal........................... Amer. Med. Assoc., Jour.
Cambridge Philosophical Society.
    Proceedings....................... Cambridge Philosoph. Soc., Proc.
Deutsche Chemische Gesellschaft.
    Berichte.......................... Deut. Chem. Gesellsch., Ber.
National Academy of Sciences.
    Proceedings....................... Nat. Acad. Sci., Proc.
Torrey Botanical Club.
    Bulletin.......................... Torrey Bot. Club, Bull.
U. S. Bureau of Standards.
    Bulletin.......................... U. S. Bur. Standards, Bull.
U. S. Bureau of Standards.
    Journal of Research............... U. S. Bur. Standards, Jour. Res.
U. S. Bureau of Standards.
    Technical Papers.................. U. S. Bur. Standards, Tech. Papers

(d) "Learned societies and academies of Europe, other than English, with names beginning with an adjective denoting royal privilege are entered under the first word following the adjective. These adjectives, Kaiserlich, Königlich, Reale, Imperiale, etc., are abbreviated to K., R., I., etc., and are disregarded in the arrangement."

(e) "Colleges and universities having a geographical designation are entered under the name of the city, state, or country contained in the title."

(f) "Observatories, botanical and zoölogical gardens, etc., not having a distinctive name, are entered under the name of the place in which they are located, unless they are affiliated with a university, in which case they are entered under the name of the university."

2. Using the Original Language. The vernacular should be used, not a translation. Just as one looks for a book by Felix Klein under *Klein,* not under *Small* or *Little,* so must one look for the Polish academy under its Polish name and use it in printed citations even though one cannot pronounce it.

Use Lund. Observ. Meddel.; *not* Contributions of the Observatory of Lund.
Do not use Acad. *for* Akad. *or* Accad.

3. Avoiding Extreme Brevity. Abbreviation of words—particularly of the first word—should not be carried too far.

In such cases as "Ann.," "Biol.," or "Geol.," when this is the first word, either the rest of the title should be written so as to leave no doubt as to the language of the title, or else the first word should be written in full. "Anales," "Annalen," "Annals," and "Annual" are far apart in a large catalogue. "Ann. d. Phys." might be either French or German.

Use Amer. Chem. Soc., Jour.; *not* J. A. C. S.
Use Arch. f. Tech. Mess.; *not* A. T. M.
Use Optic. Soc. Amer., Jour.; *not* J. O. S. A.

4. *Including All Important Words*. All important words should usually be included in the abbreviated title. The abbreviations consisting only of "Ber.," "Ann.," "Compt. Rend.," etc., in the journals published by the American Chemical Society are too short to be understood without reference to the key given by *Chemical Abstracts* in its "List of Periodicals Abstracted."

5. *Omitting Articles and Prepositions*. Articles and prepositions may be omitted when their omission does not lead to obscurity.

But use Soc. de Biol., *not* Soc. Biol., *for* Société de Biologie, *to avoid confusion with* Société Biologique.

6. *Names of Places and Persons*. Names of places and persons should not be abbreviated.

Cambridge Philosoph. Soc., Proc.
Franklin Inst., Jour.
Liebig's Ann. d. Chem.
Inst. Pasteur, Bull.

But note that Amer. (for American), Brit. (for British), Deut. (for Deutsche), and U. S. (for United States) are commonly accepted.

7. *Editor's Name*. An editor's name should be avoided unless it is in the official title.

Use Ann. d. Physik, ser. 2, *or* Ann. d. Physik [2]; *not* Poggendorff's Annalen.
But Pflüger's Archiv *is correct*.

8. *Part, Section, or Division*. When publications of an institution are organized in parts, the section or division designation should be included.

Preuss. Akad. d. Wissens., Phys.-Math. Kl., Sitz. Ber.
Akad. d. Wissens., Wien, Sitz. Ber. 2A

9. Series Number. The series number should always be given, in addition to the volume number and year number, in case the set is numbered in series.

Philosoph. Mag., ser. 7, or Philosoph. Mag. [7].

10. List of Abbreviations. The following list shows some common abbreviations of words in the names of periodical publications:

Abstracts	Absts.	Bulletin	Bull.
Academy	Acad.	Bureau	Bur.
Agricultural	Agric.	Centralblatt	Centralbl.
American	Amer.	Chemical	Chem.
Anales	An.	Chemie	Chemie
Analytical	Analyt.	Chemistry	Chem.
Anatomical	Anat.	Chimie	Chimie
Annalen	Ann.	Clinical	Clin.
Annals	Ann.	Comptes	Compt.
Annual	Ann.	Contributions	Contr.
Anthropological	Anthropol.	der	d.
Anzeiger	Anz.	Deutsche	Deut.
Association	Assoc.	Diseases	Dis.
Archiv	Arch.	Ecology	Ecol.
Archives	Arch.	Economics	Econ.
Archivio	Arch.	Edition	Ed.
Astronomical	Astron.	Electric	Elec.
Bacteriology	Bacteriol.	Engineering	Engineer.
Bakteriologie	Bakteriol.	Ergebnisse	Ergebn.
Beiträge	Beitr.	Ethnology	Ethnol.
Berichte	Ber.	Experiment	Exper.
Biochemical	Biochem.	Experimental	Exper.
Biological	Biol.	Experimentale	Exper.
Biologie	Biol.	für	f.
Biologique	Biol.	Gazette	Gaz.
Botanical	Bot.	Gazzetta	Gazz.
Botanik	Bot.	General	Gen.
Botanisches	Bot.	Genetics	Genet.
Botany	Bot.	Geographical	Geogr.
British	Brit.	Geological	Geol.

Geologische	Geol.		Psychological	Psychol.
Gesellschaft	Gesellsch.		Psychology	Psychol.
History	Hist.		Publication	Pub.
Industry	Indus.		Quarterly	Quart.
Institute	Inst.		Record	Rec.
International	Internat.		Rendus	Rend.
Jahrbuch	Jahrb.		Report	Rept.
Jahresbericht	Jahresb.		Research	Res.
Journal	Jour.		Review	Rev.
Magazine	Mag.		Revue	Rev.
Mathematics	Math.		Rivista	Riv.
Mechanical	Mech.		Royal	Roy.
Medical	Med.		Science	Sci.
Medicine	Med.		Scientific	Scient.
Monographs	Monogr.		Scienze	Sci.
Monthly	Month.		Service	Serv.
Morphologisches	Morphol.		Society	Soc.
Morphology	Morphol.		Station	Sta.
National	Nat.		Surgery	Surg.
Natural	Nat.		Survey	Surv.
Neurology	Neurol.		Technology	Technol.
Paleontologie	Paleontol.		Therapeutics	Therap.
Paleontology	Paleontol.		Transactions	Trans.
Pathology	Pathol.		Tropical	Trop.
Pharmacology	Pharmacol.		United States	U. S.
Philosophical	Philosoph.		und	u.
Physical	Phys.		Verhandlungen	Verhandl.
Physik	Physik		Zeitschrift	Ztschr.
Physikalische	Physikal.		Zeitung	Ztg.
Physiological	Physiol.		Zentralblatt	Zentralbl.
Physique	Physique		Zoologie	Zool.
Political	Polit.		Zoologischer	Zool.
Proceedings	Proc.		Zoology	Zool.
Protistenkunde	Protistenk.			

ABSTRACTS AND QUOTATIONS

ABSTRACTS

1. Form. Reference to a cited publication should usually
be made in the form of an indirect quotation or a brief

abstract that summarizes the discussion presented in the original publication.

2. *Credit.* Always give credit for ideas taken directly from any publication.

3. *Citation.* A citation of each article mentioned must appear in your literature cited or in a footnote.

4. *Punctuation.* Indirect quotations should not be enclosed in quotation marks.

QUOTATIONS

1. *Permissions.* Written permission must be obtained from the copyright owner before printing or otherwise reproducing material from a copyrighted publication. If a publisher or organization holds the copyright, it is important as a matter of professional courtesy to obtain also the author's permission. Always secure permission from the original publication—not from one that has reproduced the material. In books the name of the copyright owner will be found on the title page or its reverse. In periodicals it will be found on the title page, the first page of text, under the title heading, or on the front cover.

When writing to the copyright owner, tell how you wish to use the material and identify it clearly. For a book, give the author, title, edition number, year of publication, and page number; identify illustrations or tables by number, and text material by beginning and ending phrases. For a journal, give the journal title, volume and page numbers, article title and author's name; identify illustrations, tables, and text material in the same manner as for a book.

2. *Form.* When direct quotations are needed, they should be enclosed in quotation marks or printed in small type. They should reproduce the exact words of the original publication, including all details of spelling, diacriti-

cal markings, capitalization, and punctuation. Corrections
or remarks inserted by the one who quotes must be placed
in square brackets [ ]. Omissions must be indicated, by a
series of four widely spaced periods. The author should
carefully compare the typewritten copy with the original
printed matter each time the manuscript is copied.

3. Short Quotations. A short quotation should not appear as a separate paragraph. It should be enclosed in
quotation marks and included in a paragraph of your
manuscript.

4. Long Quotations. A quotation of more than five or
six lines should be given as a separate paragraph. Quotation marks are omitted, and the quotation will be printed
in smaller type than that used for the text.

Each quotation that is to be printed in small type should
be typewritten upon one or more separate sheets of paper
that are numbered consecutively with the text pages but
bear no text material. The method of preparing the copy
is as follows: When the place is reached for a long quotation, remove the text sheet from the typewriter and begin
the quotation (double-spaced) upon a separate sheet numbered as a new page. Finish typewriting the quotation,
using as many sheets as necessary and numbering them as
manuscript pages. Then put a new sheet of paper in the
typewriter and continue with the text.

This method allows the article to be composed economically on the typesetting machine, which will not set two
different sizes of type in one operation. If the manuscript
is not prepared in this way, the compositor must handle all
of the copy twice and needlessly waste valuable time.

It is essential to mark clearly the sheets bearing quotations. This may be done by writing the word "Quotation,"
enclosed in a circle, in the upper left-hand corner and

drawing a light pencil line down the full left margin of the quoted material.

5. Quotation Within a Quotation. Use single quotation marks for a quotation within a quotation.

ACKNOWLEDGMENTS

Acknowledgments of help received from others should be made with simplicity and tact. An effusive acknowledgment may be very embarrassing to your critic or adviser. It is fitting, of course, that mention be made of suggestions, criticisms, or other forms of help that you have received, but this should be done in an appropriate way.

The form of acknowledgment and its place in the paper should be determined by the usual practice in the journal in which your article is to be published. Acknowledgment may be made by a brief statement appearing in a footnote to the title of the article. The form for a thesis may be "Prepared in the Department of —————————, under the direction of Professor ——————————." If persons other than the adviser have helped, mention of the fact may be made in the form of footnotes in the parts of the paper concerned. Another suitable place for acknowledgments is in the introduction to the paper. Some journals put the acknowledgments at the end of the paper, just before the literature cited.

ANALYTICAL TABLE OF CONTENTS

1. Analytical Outline. Before a manuscript is offered for publication, an analytical outline, or table of contents, should be prepared. Although the outline will not be printed, it has two important uses: (*a*) It aids you in making the final revisions of your paper, especially in preparing correct headlines. (*b*) It helps anyone who reads your manuscript with the object of criticizing it.

2. *Form.* The outline that follows will serve as an example of an analytical table of contents. The ranks of the headings for the various divisions of an article should be indicated in the table of contents by graded indentations. Note that the principal divisions are begun flush with the left-hand edge of the writing, the subdivisions of the principal divisions are indented 5 spaces on the typewriter, and smaller subdivisions are indented 10 spaces.

Indicate properly the comparative ranks of the topics. If two topics are logically coordinate, do not make one topic subordinate to the other. On the other hand, if one topic is logically subordinate to another, do not give them equal rank.

[Example of analytical table of contents]

CONTENTS

HEADINGS IN THE TEXT

1. Styles of Headings. The analytical table of contents, described in the preceding section, should be used as the basis for making any desirable revisions of the headings that appear in the text and for indicating the rank of the headings. Excessive subdivision of the text by means of headings should be avoided since it confuses rather than aids the reader.

The heading "Abstract" is omitted if the abstract is to be printed in distinctive type, and the heading "Introduction" is usually omitted. The heading "Literature Cited" is conveniently printed in full capitals of the smaller type used for the citations.

In the preparation of books, it is often necessary to use special styles of headings. For textbooks, boldfaced headings (indicated by underscoring with a wavy line) are generally preferred because they make the topics stand out distinctly.

Most journals, for reasons of economy and appearance, use small capitals and italics of the text type for headings. Two systems of graded headings that are in common use are described below. Both are pleasing to the eye, and economical because composed with the text in one operation.

2. Journals Using Center Headings and Paragraph Headings. Many journals, especially those with only one column of type on the page, use both center headings and paragraph side headings, as shown by the following outline:

Main Headings: CENTER HEADINGS IN SMALL CAPS.
Subheads: *Center headings in lower-case italics.*
Secondary Subheads: *Paragraph side heads in lower-case italics.*

The main headings should be typewritten in capitals as center headings. It is better not to underline them on the typewriter. The editor will underscore them with two lines to indicate small caps, or will write "s.c." in the margin.

The subheads should be typewritten as center headings in lower-case letters and underscored with a single line to indicate italics. The first word and proper names should begin with capital initials. If only two ranks of headings are needed, the center subheads should be omitted, and the side heads described below should be employed instead.

The secondary subheads should be indented as paragraph side heads and underlined to indicate italics. Only the first word and proper names should have capital initials. A period (or in some journals a period and a dash) should follow the side head, and the text of the paragraph should begin on the same line.

3. Journals Using Paragraph Side Heads Only. Many journals with a two-column format use paragraph side heads only, as outlined below:

Main Headings: PARAGRAPH SIDE HEADS IN SMALL CAPS.—
Subheads: *Paragraph side heads in lower-case italics.*—
Secondary Subheads: (1) Paragraph side heads in lower-case roman.—

Each of the main headings should be indented as a paragraph, typewritten in capitals, and followed by a period and two hyphens (em dash). The editor may mark the heading to indicate that only the first letter is to be printed as a full capital (three underlines) and the rest as small capitals (two underlines) even though proper nouns are included. Or he may write a marginal note to this effect. The main heading is followed on the same line by a subhead (see below) or by the text of the paragraph.

Each of the subheads should be indented as a paragraph (unless it follows a main heading), typewritten in lower-case letters, underlined for italics, and followed by a period and two hyphens (em dash). The first word and proper names should have capital initials.

Secondary subheads should be avoided wherever possible. But if required, they should be preceded by Arabic numerals in parentheses and typewritten in the same manner as the subheads, but without underscoring.

Chapter 4

TABLES

1. Importance. The first step in the analysis of experimental data is to arrange them in the form of tables. This part of the work may require a great deal of study before the best scheme for bringing out relations is found. Two general types of tables may be needed: (*a*) those which contain the original data, including actual observations and measurements, and (*b*) those which contain derived data, bringing out special points and conclusions. For a discussion of various kinds of tables, see Worthing and Geffner (1943). A large part of the work of interpretation of the data will have been completed when well-arranged tables have been made.

2. Unity. Each table should be a unit. A table is a short-cut means of presenting facts to the reader, and a table (like a sentence, paragraph, or article) should present one subject with distinctness. Do not attempt to bring out in a single table several comparisons of very different kinds. Very large tables are likely to be confusing.

3. Clearness. The form of the table should be arranged to secure greatest clearness. For each kind of comparison of data, there is usually one form of table that brings out the comparison most clearly and systematically. In addition to the absolute figures representing original observations, the table may include percentages, ratios, totals, averages, etc.; the latter are often of great value in making comparisons.

4. Accuracy. Every item in the table must be checked for correctness.

114

5. Economy. Since tables cost much more per page than text material, they should be used only when needed and

TABLE 3

Isotonic molalities and activities of water for solutions of orthophosphoric acid at 25°C. Reference standard, sulfuric acid*

$m\mathrm{H_3PO_4}$	$m\mathrm{H_2SO_4}$	a_1	$m\mathrm{H_3PO_4}$	$m\mathrm{H_2SO_4}$	a_1
1.0675	0.5973	0.9782	23.524	10.146	0.3528
1.7602	0.9694	.9633	24.854	10.619	.3289
2.9858	1.6192	.9331	28.202	11.746	.2778
4.3541	2.3082	.8962	28.768	11.861	.2732
5.1875	2.7320	.8690	28.995	12.068	.2649
9.1862	4.5628	.7367	31.044	12.536	.2470
9.5130	4.7401	.7231	31.982	12.827	.2366
10.284	5.0404	.6999	33.305	13.224	.2230
10.903	5.3309	.6774	34.099	13.414	.2167
11.070	5.4010	.6720	35.773	14.064	.1963
11.197	5.4660	.6669	37.824	14.428	.1857
11.938	5.7441	.6452	41.184	15.283	.1625
12.647	6.0190	.6237	43.035	15.733	.1518
13.550	6.4432	.5915	50.010	17.268	.1202
14.057	6.6349	.5765	57.265	18.766	.0957
15.685	7.2422	.5323	64.420	20.024	.0791
15.912	7.3646	.5237	64.659	20.064	.0786
18.034	8.1749	.4677	74.726	21.660	.0627
19.006	8.5554	.4432	120.56	27.01	
20.708	9.1841	.4084	134.00	29.37	
21.327	9.3940	.3937	217.4	33.69	
21.608	9.5043	.3874			

* Data from: Elmore, K. L., C. M. Mason, and J. H. Christensen. 1946. Activity of orthophosphoric acid in aqueous solution at 25° from vapor pressure measurements. Amer. Chem. Soc., Jour. 68: 2528–2532.

should not be made unnecessarily large. For a two-column page, they should be designed, if possible, to fit within a single column. Abbreviations should be used to keep the column heads of the table small. A column should not be devoted to only one or two entries, to a repetition of the same entry, or to data that may be easily calculated from data in another column. Such cases can usually be cared for in footnotes or in notes following the title.

TABLE 4

Distances progressed at different times by three strains of Neurospora crassa growing on agar medium containing a limiting concentration of l(+)leucine (0.0075 mg/ml) *

TIME	DISTANCE COVERED		
	Prototrophic f_1 (adapted)	Heterokaryon 1	Leucineless
days	mm	mm	mm
0.00	0	0	0
0.45	40	22	
0.97	89	45	
1.00			51
1.47	140	72	
1.52			73
1.88	182	95	
1.98			98
2.47			123
2.51	246	122	
3.03			147
3.10		148	

* Data from: Ryan, F. J., and J. Lederberg. 1946. Reverse mutation and adaptation in leucineless *Neurospora*. Nat. Acad. Sci., Proc. 32: 163-173.

6. Size. The table must be compiled so as to fit the page of the publication. On a two-column page, tables may occupy a single column, or, if necessary, the full width of the page.

The space that a printed table will require may be estimated by means of character counts (page 62). As an approximation, assume that in the 8-point type used for the body of the table there are 17 characters per horizontal inch and 7 lines per vertical inch, and in the 6-point type used for the box heads there are 21 characters per horizontal inch. Allowance must be made for space between columns equal to at least 2 characters.

When large tables are required, the method of handling

TABLE 5

Mean number of aleurone dots on seeds with three a_1 genes and seeds with two a_1 genes from eight ears of crosses $a_1a_1DtDt \times a_1a_1^Pdtdt$ or $a_1a_1dtdt \times a_1a_1^PDtDt$

PEDIGREE	I. MEAN NO. DOTS ON $a_1a_1a_1Dt$ CLASS	NO. OF SEEDS IN CLASS	II. MEAN NO. DOTS ON $a_1a_1a_1^PDt$ CLASS	NO. OF SEEDS IN CLASS	RATIO† I:II
2676 (1) × 2511a	15.2	46	10.0	48	1.52
2500 × 2511b	9.0	32	6.2	49	1.45
4345 (4) × 4344 (1)	10.2	152	6.1	172	1.67
4345 (6) × 4344 (7)	7.4	61	5.0	62	1.48
4345 (8) × 4344 (1)	5.6	70	4.2	76	1.33
4345 (9) × 4344 (2)	11.5	16	9.6	19	1.20
4345 (11) × 4344 (2)	3.5	39	2.6	42	1.35
4345 (12) × 4344 (8)	2.9	20	1.4	27	2.07

* Data from: Rhoades, M. M. 1938. Effect of the *Dt* gene on the mutability of the a_1 allele in maize. Genetics 23: 377–397.

† Mean observed ratio, 1.51; theoretical, 1.50.

them should be left to the judgment of the printer. If a table is too large to come within the width of the page, it may be possible to set it lengthwise on the page. If it will fit neither crosswise nor lengthwise, then it may be possible to keep it within bounds by setting it in 6-point type, the smallest size used for book and periodical work. If this method fails, the table may be spread across two facing pages.

A folder should not be used unless it is absolutely unavoidable. Folders not only are very costly, but are unwieldy for the reader and are likely to be torn when handled in the library.

7. Large Tables in Manuscript. If a table requires a larger sheet than that used for the text of the manuscript,

TABLE 6

*Average dry weight of tops and selenium content of maize grown in culture solutions containing various concentrations of sodium selenite or seleniferous Astragalus extract**

SELENIUM IN CULTURE SOLUTION	AVE. DRY WT. OF TOPS		SELENIUM CONTENT	
	Sodium selenite	Astragalus extract	Sodium selenite	Astragalus extract
ppm	*grams*	*grams*	*ppm*	*ppm*
0	1.36	1.36	0	0
1	1.25	1.28	49	566
2	1.06	0.94	94	1192
5	0.68	0.68	103	1912
10	0.42	0.37	121	2535
20	0.19	0.20	235	3150

* Data from: Trelease, S. F., and S. S. Greenfield. 1944. Influence of plant extracts, proteins, and amino acids on the accumulation of selenium in plants. Amer. Jour. Bot. 31: 630–638.

the sheet may be folded and inserted in place as one of the manuscript pages.

8. Each Table by Itself on Separate Page. Each individual table should be typewritten on a separate sheet of paper, without any of the text on the same page. This is necessary to facilitate typesetting.

When the place for a table is reached in typewriting a manuscript, the text sheet should be removed from the typewriter (no matter where the typewriting ends), and a new sheet should be inserted. Only the table (including its heading and footnotes) should be written on this sheet. The text should be continued on a fresh sheet of paper.

An alternative method is to typewrite a note in the left-hand margin at the place for the table (for example,

TABLE 7

*Comparison of microbiological and chemical method for determining methionine, and effect of presence of an equal weight of carbohydrates during hydrolysis of protein**

MATERIAL ANALYZED	METHIONINE FOUND IN PROTEIN		
	Microbiological method with		Chemical method
	Leuconostoc mesenteroides	Streptococcus faecalis R	
	per cent	*per cent*	*per cent*
Beef loin†	2.52	2.52	2.45
Beef liver†	2.34	2.24	2.27
Casein‡	2.72	2.58	2.57
Casein and sucrose	2.42	2.42	2.41
Casein and arabinose	2.49	2.55	2.46
Casein and starch	2.45	2.46	2.42

* Data from: Lyman, C. M., O. Moseley, B. Butler, S. Wood, and F. Hale. 1946. The microbiological determination of amino acids. III. Methionine. Jour. Biol. Chem. 166: 161–171.

† Protein content calculated as nitrogen content × 6.25.

‡ Difco isoelectric casein; values not corrected for moisture and ash.

"Insert table 1"), finish typewriting the page of manuscript, and then typewrite the table by itself on the next manuscript page.

9. Open and Ruled Tables. In open tables, vertical rules are omitted and horizontal rules are either omitted entirely or only three are used—one above the column headings, another below them, and a third at the bottom of the table. Ruled tables have vertical rules and often additional horizontal rules in the column headings.

10. Style. Sample tables of a common style are shown on pages 115–119. The suggestions given here refer specifically to tables prepared according to this style. But since the style varies in different journals, the author should study the tables in the journal in which the paper is to be published and should take care to prepare the tables in the proper form.

11. Heading of Table. The tables are numbered consecutively in each article. The word "Table," capitalized and followed by an Arabic number, appears as a center heading in 8-point type. The legend, or descriptive title, is centered above the body of the table and printed in 8-point italics; only the first word and proper names have capital initials. It is typewritten double-spaced and underlined for italics.

The legend should make the table self-explanatory. It should be concise and specific, but broad enough to include all the data in the table. The important words should be placed near the beginning, so that in a series of tables the subject of each can be seen at a glance.

12. Box Heads. In the examples shown in this chapter, the box heads, at the tops of columns in a table, are printed in 6-point small caps and the secondary heads in 6-point lower-case type. (Many journals, however, use 8-point lower-case type for both primary and secondary column headings.)

13. Units of Measurement. Units of quantity are given below the line under the box heads and are printed in 6-point italics. (In open tables such units, preceded by commas, are given in the column headings.)

14. Body of Table. Columns consisting of words in the body of the table appear in ordinary type. Figure columns are aligned on the right; reading columns, on the left. Figure columns are separated from vertical rules at least an en space; decimals are aligned; figures are centered in columns. Omissions are indicated by blank spaces, and the reasons for omission of important data are explained in footnotes. The body of the table is printed in 8-point type, either on 10-point base (in the examples given in this chapter) or on 9-point base (in many journals with a two-column page).

15. Footnotes. In the examples shown in this chapter, explanatory footnotes to tables are indicated by means of standard footnote reference marks (*, †, ‡, §, etc.) placed after the words or the numbers to which the footnotes refer. Many journals, however, indicate footnotes by means of superscript lower-case letters ([a], [b], [c], etc.), placed *after* words or *before* numbers in the table. The footnotes are typewritten on the manuscript sheet bearing the table. Each footnote is preceded by its symbol or superscript letter. Some journals indent each footnote as a separate paragraph; others run them all into a single paragraph.

16. Horizontal Rules. Additional horizontal rules in the body of the table should be avoided, since they increase the cost of printing. Where a line of demarcation is necessary, it can be indicated inexpensively by a blank space.

17. Spacing. The figure columns are cast to accommodate the widths of the figure entries or wording of the box heads, and to keep equal spaces between the vertical rules. The balance of the space may be put in the first column (as in table 7) or in reading columns. In a very large table, grouping the horizontal lines of figures in groups of four lines, by a double space, makes the table easier to read.

18. References in Text. References to tables are made by number. (For example: The data of the second experiment are presented in table 3.)

Chapter 5

ILLUSTRATIONS

Illustrations form an integral part of the concise and effective presentation of scientific and technical material. They serve as a short-cut means of presenting descriptive matter and of showing relations among data. Illustrations attract the attention and interest of the reader. They quickly give him information that he would otherwise have to obtain from long verbal explanations, and they give him a clear conception of objects or relations that are too complex to be adequately described in words. If the paper is suitably illustrated, the text may be largely devoted to comparisons, inferences, and discussions of principles.

Photographs and drawings are especially important in the descriptive phases of science. Diagrams of apparatus and graphs of data are mainly required in the experimental and quantitative phases.

The first impression a reader gets of an article is greatly influenced by the appearance of the illustrations. He is likely to receive an unfavorable impression of the whole article if the illustrations are poor. But he is attracted to the article if the illustrations are clear, artistic, and informative.

It is very important, therefore, to devote much time and study to the planning and preparation of illustrative material. Each illustration should be a unit, presenting a single subject as clearly and distinctly as possible. Special attention should be given to uniformity in style, tone, and lettering. For putting drawings and graphs into final form, the services of a professional artist or draftsman may be needed.

CORRECT PROPORTIONS

An illustration with proportions approximating 1 by 1½ is most pleasing to the eye. The appearance of the page is usually best and the printer's work is facilitated if the illustration has the same width as the type column or type page. But a simple drawing should not be printed on a grotesquely large scale in order to have it fill this width.

It is necessary to plan the illustrations in conformity to the requirements of the journal in which they are to be published. For a journal printed in two columns, illustrations should usually be designed to occupy the width of a single column, and should therefore be tall rather than wide. To be readily seen, the essential part of an illustration must usually be shown on as large a scale as possible within the area allowed by the width of the type column or type page. This is especially important when the illustration is limited to the width of a single column of a two-column page. Only by careful planning, proportioning, and exclusion of blank background and all extraneous matter can the informative part of the illustration be shown on a sufficiently large scale.

1. Graphical Method. Figure 1 shows a convenient graphical method of obtaining correct proportions for an illustration that, together with its legend, will occupy a full column or page.

On a large sheet of Bristol board, or of stiff white cardboard on which a group of drawings or photographs is to be mounted, draw in pencil a rectangle, *ABCD,* that is the exact size of the desired reproduction. This rectangle is the same in width as the type column or page, but it is

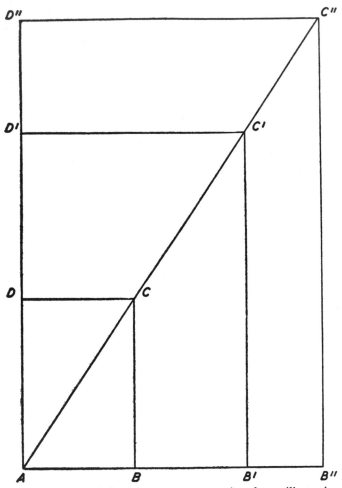

Fig. 1. Method of obtaining correct proportions for an illustration.

enough shorter to allow space for the legend that will be printed below the illustration. (Some journals allow an illustration to occupy the full height of the type page, and they print the legend at the foot of the other column on the same page or at the foot of the facing page.)

The space for the printed legend may be determined as follows: Count the total number of typewritten characters (including blank spaces) in the legend. Divide this number by the number of characters in a printed line (e.g., by 55 for a 3-inch line or by 115 for a $6\frac{3}{16}$-inch line) in order to obtain the number of printed lines. Then obtain the height of the printed legend as follows: (*a*) If there are *7 lines* per vertical inch, allow $\frac{3}{16}$ inch for the first line and $\frac{1}{7}$ inch for each succeeding line. (*b*) If there are *8 lines* per vertical inch, allow $\frac{3}{16}$ inch for the first line and $\frac{1}{8}$ inch for each succeeding line.

Having drawn the rectangle *ABCD* the size of the reproduction, extend the diagonal *AC* as far as you wish on the Bristol board or cardboard. Any point on the diagonal will determine a rectangle that has the correct proportions of width and height. For example, the point *C'* determines the correct rectangle *AB'C'D'*, and point *C"* determines the correct rectangle *AB"C"D"*.

If the reproduction is to occupy only a part of the width or height of the type column or page, a similar procedure is used. The original, small rectangle is always made the exact size of the intended reproduction.

2. Checking Completed Illustration. A completed illustration may easily be checked to find out whether its height or its width must determine the reduction. Cut a sheet of paper to the size of the type column or type page *minus* the height of the legend. Place it on the illustration in the position of *ABCD* in figure 1, and lay a long ruler on the illustration so as to extend the diagonal

AC. Then: (*a*) if the ruler intersects the *side* of the illustration, the *height* must be reduced to that of the type column or page *minus* the height of the legend; or (*b*) if the ruler intersects the *top* of the illustration, the *width* must be reduced to that of the type column or page.

DRAWINGS[12]

1. Methods of Reproduction. Drawings are usually reproduced by means of zinc etchings, or by the more expensive halftone and photogelatin processes. Before the original drawing is made, it is necessary to know what method of reproduction will be used, how much the drawing will be reduced, and on what type of paper it will be printed. Where possible, drawings should be prepared for reproduction by the relatively inexpensive zinc etchings. Halftones cost about twice as much, and photogelatin prints cost from 5 to 15 times as much as zinc etchings.

2. Text Figures and Plates. Drawings are commonly used as text figures, but sometimes as plates. Text figures are printed on the same paper as the text and may have text material above or below them. Plates are used in a few journals for halftone or photogelatin reproductions; they are often printed on special paper as separate pages and inserted at the end of the article. Most journals, however, use the same paper for all illustrations and treat them as text figures, even if they are grouped so as to occupy a full page.

3. Ink and Paper. Undiluted black waterproof India ink should be used in preparing drawings for reproduction by zinc etchings. Drawings should first be completed

[12] Useful books on scientific and technical drawings are those of the Higgins Ink Co. (1948, 1953), Ridgway (1938), Staniland (1953), and the Wistar Institute (1934).

in pencil and later inked in. Pure white three-ply Bristol board is used for most drawings. If a drawing has been made on thin paper, it should be transferred to Bristol board by blackening the back with a soft pencil and tracing over the drawing with a hard pencil.

For diagrams of apparatus, coordinate paper ruled with nonreproducible blue lines (see page 132) is sometimes convenient because the lines serve as guides. For certain types of maps and charts, a method of scribing on coated Stabilene Film (Keuffel and Esser Co.) offers advantages over conventional methods of drawing.

4. Size of Drawing. The width or length of the original drawing should be from 2 to 3 times that of the reproduction. Slight inaccuracies in lines become invisible when reduced. Standard enlargements should be used for drawings in the same article, so as to insure ready comparability.

5. Shading for Zinc Etchings. For reproduction by zinc etchings, any shading that is desired should be done by means of black dots or lines made with undiluted India ink. Darker shades are best obtained by putting the dots closer together, rather than by increasing the size of the individual dots. Very fine lines or extremely small dots cannot be reproduced by this process. They will be lost unless they are clear and distinct. If they are placed too close together, they will blur and appear as solid blotches when the drawing is reduced in size. There is a greater tendency for blurring to occur when the drawing is printed on a soft or rough-surfaced paper than when it is printed on a hard or glossy-surfaced paper. The shading should be kept rather open. A light drawing is more attractive than a dark one. Many of the best drawings are mere outlines, made with very few, carefully chosen lines. Elaborate drawings are rarely necessary.

Rather delicate shading can be obtained in the zinc etching if the size and spacing of the dots are properly adjusted to the degree of reduction and to the type of paper used in printing. Sample drawings that have given good results should be studied. Each laboratory should preserve original drawings, together with their reproductions, for the guidance of its workers. Examination of the drawing through a reducing lens is helpful, but the final test is the reproduction itself. In case of doubt, it will pay to have a zinc etching made from one of the drawings before proceeding with the preparation of the rest.

Copper etchings, which cost about twice as much as zinc etchings, permit the reproduction of somewhat finer lines and smaller dots.

6. Shading for Halftones. If it is essential to show gray tones or extremely fine details, very delicate shading may be done with several dilutions of India ink, and the drawings can be reproduced by the halftone process or the photogelatin process. Wash and brush drawings, made with water color or diluted India ink, can be reproduced in the same way. India ink, once applied, cannot be lightened; but water color can be partially removed by strokes with a brush dipped in water. Pure white Bristol board should be used—never cardboard with a cream or yellow tint.

7. Economy in Grouping. As many as possible of the illustrations should be grouped and mounted close together on heavy white cardboard, so that they may be reproduced as a single cut. Grouping is economical because the photoengraver's charge for one-half page is about three-quarters of that for a full page, and his minimum charge, for small figures, is about one-half the charge for a full page.

8. Grouping for Zinc Etchings. For reproduction by a zinc etching, the separate drawings on Bristol board should be trimmed (to within about $\frac{1}{4}$ inch of the perimeter of each), and carefully arranged within a rectangle of proper size and proportions drawn in pencil on a sheet of stiff white cardboard (Bainbridge board no. 80, about $\frac{1}{16}$ inch thick). The trimmed edges of the drawings will not show in the zinc etching. When the best arrangement has been obtained, the figures should be fastened to the cardboard with rubber cement.

Figure numbers and explanatory letters should be put in proper places. Capital letters (for major parts or units) and lower-case letters (for minor parts or units) may be used to designate points, lines, objects, etc., in drawings. Care should be taken to have the numbers and letters set straight and duly separated from the drawings. They should be neat. Heavy, blackfaced characters should never be used. The size of the characters should be such that the figure numbers will be $\frac{3}{32}$ of an inch high and the capital letters $\frac{1}{16}$ of an inch high when reduced in the reproduction. If the lower-case letters are used, they should be $\frac{1}{16}$ inch, and the capitals and numerals used with them should be $\frac{3}{32}$ inch. It is often best to paste on printed characters, which may usually be obtained in various sizes from the editor of the journal in which your paper will be published. Numbers and letters or full labels may be put on the drawing in India ink with the aid of a Wrico or Leroy lettering guide (see table 8). It is also possible to paste in place labels that have been composed by a printer. Labels typewritten with an electric machine having proportional spacing are satisfactory because they resemble printing.

9. Grouping for Halftones. Special precautions need to be taken in trimming and mounting a group of drawings for halftone reproduction. The pieces of Bristol board bearing the drawings should be cut into rectangles or polygons that fit together perfectly, so as to cover completely the white cardboard (Bainbridge board no. 80) on which they are mounted. Obtaining a suitable layout may be facilitated by making trials with tracing-paper outlines of the drawings. The size and proportions of the whole group should be determined as described on page 123. For mounting the figures, rubber cement is most convenient because it allows the position of each figure to be adjusted accurately. The photoengraver will cut thin white lines to separate the individual rectangles or polygons. If mounting is done improperly, it will be necessary to have the whole background routed out or to use highlight halftones, thus doubling or trebling the cost of reproduction. Failure to observe proper precautions in trimming and mounting is the most common cause of untidy appearance and unnecessary expense in halftone reproduction.

Figure numbers and explanatory letters may be put directly on the original drawings with the aid of Wrico or Leroy lettering guides (see table 8). Printed characters may be attached with rubber cement. All the slips of paper bearing them should be the same size, trimmed square, and set straight.

DRAWINGS FROM PHOTOGRAPHS[13]

1. Advantages. A line drawing may be made from a photograph. For illustrating a piece of scientific apparatus, such a drawing may be much better than a photograph, because the drawing shows only the points that are essential and omits unnecessary and confusing details. The drawing may be made from an ordinary view, an exploded view, a phantom view by double exposure, or successive views by multiple exposure. Hidden parts can be shown by cut-away sections. A very natural per-

[13] Mr. John W. McFarlane, of the Eastman Kodak Co., has helped in the preparation of this section.

spective may easily be obtained in a drawing based on a photograph.

The technique is simpler and quicker than the free-hand method. The drawing may be used to illustrate any type of subject matter, and it may be a simple outline or a realistic picture. The final result is limited only by the skill of the draftsman. An excellent drawing may be made from a relatively poor photograph. The line drawing is reproduced by means of a zinc etching. It costs less to publish than a halftone, and the quality of the reproduction can be better predicted. If a small number of copies of a report is needed, a line drawing can be reproduced well by Photostat copies or other photographic methods.

2. Method. The simplest way to prepare such a drawing is to trace it from a print made on 8 x 10 inch or larger single-weight paper. A sheet of tracing vellum is placed over the print, and a light is put below it for transillumination (see description of table for tracing, page 141). The desired lines are then traced in pencil on the vellum. Details that are not wanted are not traced. It is easy to study the progress of the drawing, and to disassociate it from the photograph, merely by turning off the light and observing the drawing by reflected light alone. Any shading that is desired may be put in with stippling or hatching, as in making ordinary pen drawings. Measurements, explanatory letters, and labels may be inserted. After the drawing has been completed in pencil, it is inked in. The width of the ink lines must be properly related to the reproduced size of the drawing.

If much work of this kind is to be done, a semiautomatic method is provided by the Kodak Tone-Line process. This involves making high-contrast negatives and positives, but it requires a minimum of hand drawing.

GRAPHS[14]

Graphs are designed to portray relations existing among data. They must be accurate, and they should also be clear. Since the ease with which the relations may be seen depends upon unity, balance, and other features of good composition, graphs should be constructed so as to be pleasing to the eye. The suggestions given here are intended as guides to the achievement of effective presentation. Figures 2 to 6 exemplify some of these suggestions. Models of good style are to be found in many of the scientific and technical journals. Good examples of histograms, bar graphs, pie graphs, and other special forms may also be found.

Uniformity in style may be achieved if a good set of specifications, such as that given on page 144, is adopted for all the graphs in a publication. Adoption of these specifications will not make your graphs look like everyone else's. It will make them stand out as distinctly superior to most that appear in the journals.

1. Paper and Ink. Curves and other types of graphs are reproduced as zinc or copper etchings. The simplest method of preparing a graph is to draw it first in pencil and then in black waterproof India ink on a sheet of coordinate paper ruled with nonreproducible light-blue lines. In the photoengraving process the blue lines are filtered out, and only the black ink lines remain. Paper for graphs should be of good enough quality to take India ink well and to permit corrections to be made by erasing and redrawing.

A suitable transparent coordinate paper, made of high-

[14] This section follows many of the recommendations of the American Standards Association (1943).

A comprehensive discussion of the representation of data by tables, graphs, and equations is given by Worthing and Geffner (1943).

grade tracing paper ruled into squares with 10 nonreproducible blue lines per ½ inch, may be obtained in sheets of 8½ x 11 inches (Eugene Dietzgen Co. no. 340AN-20 or Keuffel and Esser Co. no. 195GL-111) or in sheets of any desired length cut from continuous rolls 36 inches wide. Inexpensive photographic copies of graphs that have been drawn on transparent paper may be made by a direct printing process, such as Ozalid.

A satisfactory opaque paper comes in sheets 17 x 22 inches ruled into squares with 10 nonreproducible blue lines per inch (Eugene Dietzgen Co. no. 360 or Keuffel and Esser Co. no. N331). Copies may be made by the inexpensive Verifax process or by the more expensive Photostat process, the latter giving copies of reduced or enlarged size if wanted.

Coordinate paper ruled into equal squares is suitable for most types of graphs, including simple logarithmic graphs. For some purposes, paper with semi-log or log-log rulings is more convenient.

Papers ruled with dark-blue, green, yellow, orange, red or black lines are unsatisfactory unless the graph is to be transferred in India ink to tracing paper (or cloth). (Some workers actually prefer to transfer all graphs to tracing paper, after having drawn them first in pencil on thin paper ruled with green or orange lines.)

2. Size and Proportions. The general suggestions that have been made regarding the size and proportions of various illustrations apply to graphs (see page 123). The printed reproduction should have a large enough scale to show essential details and accommodate legible numbers and labels.

For convenience in mailing the original graph and copies with the manuscript, it is desirable to have the graph on a sheet of standard manuscript size, 8½ x 11 inches. A clear margin of at least an inch should be left around the graph. If a larger sheet is used, the original graph should be kept flat, but copies may be folded to fit in the manuscript.

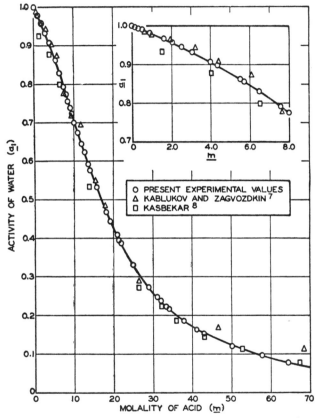

FIG. 2. Activity of water over orthophosphoric acid solutions at 25°C. (*Elmore, K. L., C. M. Mason, and J. H. Christensen. 1946. Amer. Chem. Soc., Jour. 68: 2528–2532.*)

Plotted from data of table 3, page 115.

3. Planning the Graphs. A plan should be made in advance for all the graphs in an article. For a journal with a two-column page, most graphs should be designed to occupy the width of a single column (usually 3 in.). Care must be taken to present the essential information on as large a scale as possible within the area allowed by this width. Only in exceptional cases is it necessary to use the full two-column width (usually $6\frac{3}{16}$ in.). A part or all of the height of the type page (usually $8\frac{5}{8}$ in.) may be utilized for the graphs and their legends.

For economy in photoengraving, as well as in the use of space in the printed article, some journals require that as many as possible of the graphs be grouped and mounted together on heavy white cardboard, for reproduction as a single cut of column or page width. It is often feasible, for example, to group two or three graphs in a column, and have the printed legends of all below the group. A good procedure is to make a tracing of the type page, and to plan a layout for all the graphs before work on any of them is begun. After the graphs have been finished, but before they are mounted, photographic copies of the individual graphs, preferably no larger than $8\frac{1}{2}$ x 11 inches, should be made for use in mailing with the manuscript to reviewers. Mounting should be postponed until the paper has been accepted for publication and the plan has been approved by the editor.

4. Enlargement of the Original Graphs. Best results are obtained by drawing all the originals to the same scale of enlargement—preferably twice the dimensions of the intended reproductions. This simplifies the necessary drawing equipment, facilitates the photoengraving process, and insures uniformity of reproduced letters, numbers, and lines.

In calculating the space to be occupied by the printed

graph, use its over-all dimensions, including the labels
and numbers along the axes.

One method of preparing an original graph that has
twice the dimensions of the intended reproduction requires some simple calculations, which are easily made
with a slide rule or calculating machine. The data are
arranged in a table that shows the calculated abscissa
and ordinate for every point to be plotted, and also the
positions of the scale markers or coordinate rulings. The
graph is then drawn in pencil on coordinate paper ruled
with nonreproducible blue lines, and is finally inked in.
The following example illustrates the method of making
the calculations:

Suppose that the original graph for figure 3 is being planned.
The original graph is to have twice the dimensions of the printed
graph, and it is to be drawn on coordinate paper ruled with 10 units
per ½ inch.

The total width of the original graph is to be 2×3, or 6 inches.
This is 6×20, or 120 units on the coordinate paper.

It is estimated that the label and scale numbers (0.120 in. high)
on the left of the grid will require 10 units of the width, and that
the last number on the horizontal scale will require 1 unit—making
a total of 11 units. The grid width can therefore be 120 minus 11,
or 109 units. The plotted abscissa of each point and of each scale
marker is obtained by multiplying the actual abscissa by 109/7, or
15.6.

The grid height of the original graph is to be 2×2.25, or 4.5
inches. This is 4.5×20, or 90 units on the paper. The plotted
ordinate of each point and of each scale marker is obtained by multiplying the actual ordinate by 90/45, or 2.00.

A photographic method avoids the necessity of calculations. The original graph is drawn in firm black lines
with a sharp pencil (of HB or F grade) to any convenient
scale (from 1 to 4 times the dimensions of the reproduction) on paper ruled with lines of any color. It is then
reduced or enlarged photographically (as by Photostat

copy) to exactly twice the dimensions of the intended reproduction. Only a negligible error is usually introduced by unequal shrinkage of the photographic print. The graph is finally transferred in India ink to high-grade tracing paper.

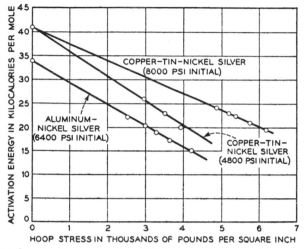

FIG. 3. Activation energy for diffusion as a function of hoop stress. (*Mason, W. P., and O. L. Anderson. 1954. Bell System Tech. Jour. 33: 1093–1110.*)

5. *Choice of Coordinates.* It is customary to plot the independent variable on the horizontal axis and the dependent variable on the vertical axis. Intervals of time are plotted on the horizontal axis.

6. *Scale of Coordinates.* The scale of coordinates should be chosen so that the printed graph will be neither crowded nor wasteful of space. The same scale should be used in a series of comparable graphs.

If the graph is to be used as a source of quantitative

data, the scale should be such as to allow the coordinates of any point to be read easily and accurately. But great precision is impossible in reproductions because of distortion of the gelatin film used in their preparation. If the graph is presented to illustrate the nature of the relation between the variables, a smaller scale may be used. Most of the graphs in scientific papers are of this type, especially when the original data are presented in tables.

It may be desirable to try to find a method of plotting that will give a straight line. If this can be done, it may give a clue to the mathematical relation between the two variables. If the ordinary graph is straight, the relation follows the linear law $(y = ax + b)$; if a log-log graph is straight, it follows the power law $(y = kx^n)$; or if a semilog graph is straight, it follows the compound-interest law $(y = Pe^{rx})$. In many cases, of course, none of these graphs is straight, and the mathematical relation between the two variables is more complex.

Some readers prefer plots made from the original data, since they find it difficult to visualize relations when the logarithms of the variables are plotted. Both types of plots may sometimes be put in the same graph, with one scale arithmetic and the other logarithmic.

The coordinates should be chosen so that the important part of the curve approximates a slope of unity—i.e., makes an angle of about 45° with horizontal axis.

For arithmetic scales, round numbers—1, 2, or 5, multiplied or divided by 1, 10, 100, etc.—and corresponding scale markers are put on coordinate scales divided in the decimal system.

For logarithmic scales, markers are ordinarily put at values of 1, 2, 4, 6, 8, 10, 20, 40, 60, 80, 100, etc.

The scale labels should be balanced near the middle

of the axes and should not be crowded too close to the numbers on the scale. The label on the vertical axis should be oriented so that it is read upward along this axis. Each label should indicate clearly (*a*) the name of the variable plotted and its symbol if one is used in the text,

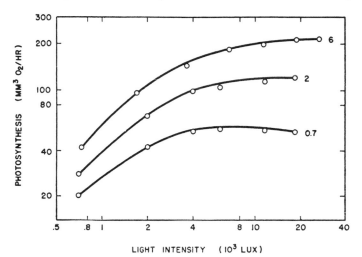

LIGHT INTENSITY (10^3 LUX)

Fig. 4. Rate of photosynthesis of *Chlorella pyrenoidosa* per 100 million cells in relation to light intensity. Cultural intensities are shown by curve labels. (*Adapted from: Winokur, M. 1948. Amer. Jour. Bot. 35: 207–214.*)

and (*b*) the unit of measurement. Suitable forms are "Pressure *p* (lb/in.²)" and "Pressure *p*, lb/in.²" But the style should conform to that of the journal in which the paper is to be published. The technical terms, symbols, and abbreviations in a graph should be in accord with those used in the text of the article.

The use of many digits in scale numbers may be avoided by using a large unit of measurement. For example, instead of using "10,000, 15,000, 20,000" and the label "Pressure (lb/in.^{2})," use "10, 15, 20" and the label "Pressure (10^3 lb/in.^{2})." In figure 3 the labels might have been shortened to "Hoop stress (10^3 lb/in.^{2})" and "Activation energy (kcal/mole)."

7. Plotting the Points. The points are plotted with a sharp pencil (of HB or F grade) as small but clear dots, and each point is surrounded by a circle or other symbol. After the points have been plotted, all should be carefully checked. Observed points should be clearly shown in the curve. But computed points, plotted from a mathematical equation, should not be shown, unless for some special reason explained in the text.

8. Drawing the Curve. If the points are widely or irregularly distributed, all that can be done is to connect them with straight lines. But where possible, a smooth curve should be drawn to represent the plotted points (fig. 2, 4). Smoothing shows relations most clearly and minimizes errors. In some cases a smoothed curve is fitted to the observed points with the aid of a theoretical equation given in the article.

In drawing a smooth curve by hand, turn the paper so that your hand is on the concave side of the curve and use light sweeping strokes with a pencil. Obtain a satisfactory curve by repeated light erasure and correction. To detect kinks or humps that require correction, hold the sheet of paper so that its edge is level with the eye, and sight with one eye along the curve. The smoothed curve need not pass through all the points. It should be a mean curve, drawn so that about half the points in a group fall on each side of it.

Some workers like to draw the smoothed curve on a piece of tracing paper fastened over the graph with drafting tape. The tracing paper is then removed and mounted with rubber cement on a piece of cardboard (such as a

5 x 8-inch index card or a filing folder), which is then cut with scissors along the curve. The cardboard template is used in drawing the curve in pencil on the original graph. A very smooth curve may be obtained in this manner.

9. Several Curves in a Graph. Several curves may be drawn in the same graph, but they should not be so numerous or crowded as to make the graph difficult to decipher. Ordinarily, not more than three or four should be drawn unless they comprise a family of well-separated curves. The curves may be distinguished, as in figure 5, by different symbols representing the points (circles, triangles, and squares) and by different kinds of lines (wide solid, narrow solid, long dash, short dash, and long and short dash). Outlined symbols are preferable to filled-in symbols, except for scatter diagrams (fig. 6).

Wherever possible, the individual curves should be designated by labels running along the curves (fig. 6) or placed horizontally near them (fig. 3). They should not be placed at the ends of the curves, where they would increase the width of the graph. If there is insufficient space for labels, the curves may be identified by a key in a balanced position in the graph (fig. 2), or by reference in the printed legend to characters indicating the curves (fig. 4, 5).

10. Tracing Graphs. It is often necessary to transfer a graph to tracing paper of high quality, such as Age-Proof Vellum of Eugene Dietzgen Co. or Albanene of Keuffel and Esser Co. (Tracing cloth is much more expensive and offers no advantage except greater durability.) This must be done if the graph has been made on coordinate paper ruled in any color other than nonreproducible blue, or if it is a photographic copy of an original in pencil.

Transillumination, though not essential, is an aid in tracing graphs. A convenient table may be constructed by cutting a rectangular opening of suitable size (17 x 22 inches) in a wooden table and mounting above the opening a countersunk sheet of ground plate-glass (20 x 25 inches). The glass may be illuminated by an incandescent

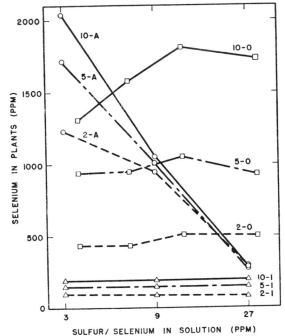

Fig. 5. Selenium accumulation by maize in relation to sulfur/ selenium ratio (in ppm) in the culture solution.

Cultures received 2, 5, and 10 ppm of selenium as selenate (*A*), selenite (*I*), or organic selenium (*O*) from an *Astragalus* extract. Sulfur supplied as sulfate. (*Trelease, S. F., and O. A. Beath. 1949. Selenium.*)

lamp in a stand resting on the floor, or by suitably placed fluorescent lamps. The graph is fastened to the glass with drafting tape. Then the tracing paper, after being adjusted over the graph for alignment and margins, is secured with drafting tape to the glass.

The graph can be transferred in India ink to the tracing paper without first making a pencil tracing. Only a few pencil guide lines are needed, such as those for axes and scale markers and for numbers and letters along the axes.

11. Inking the Graph. Inking in the graph is the part of the work that is best done by a professional draftsman. With several hours of practice, however, a beginner can acquire sufficient skill to do the work well, and he will have the satisfaction of learning to finish the graph himself. Instructions for the handling of ruling pens and other tools are given in a book by the Higgins Ink Co. (1953).

Standard specifications for an original graph having *exactly twice* the dimensions of the intended reproduction are shown in the outline on page 144. Their adoption simplifies the attainment of legibility and uniformity of all the graphs in a paper.

If a different degree of enlargement has been used for the original graph, then all symbols, numbers, letters, and line thicknesses should be changed accordingly. Uniformity in the reproductions is difficult to achieve, however, when several scales of enlargement have been used in preparing the various graphs for a paper. It is usually easiest and best in such a case to have each of the graphs reduced or enlarged photographically (by a Photostat copy) to twice its reproduced size, and then to transfer the graphs to tracing paper and use the standard specifications in inking them in.

A good order for inking the graph is: (a) data points, (b) curves, (c) lettering within the graph if full coordinate rulings are to be used, (d) axes, (e) stubs (ticks) or coordinate rulings, (f) numbers and labels along the axes.

Specifications for an original graph having exactly twice the dimensions of the intended reproduction

Wide line for the principal curve. Make with Wrico pen no. 4 or Leroy pen no. 3 (2½ points or 0.035 in.). If there are several equal curves in the same graph, they may be made with Wrico pen no. 6 or Leroy pen no. 2 (2 points or 0.026 in.). A light curve may be made with Leroy pen no. 1 (1½ points or 0.021 in.).

    (━━━━━━━━ 2-point rule.)

Medium line for numbers and capital letters made with guide giving characters 0.120 in. high, and for axes or rectangular frame. Make with Wrico pen no. 7 or Leroy pen no. 0 (1¼ points or 0.017 in.). If a lighter line is desired for the axes or frame than for the lettering, use Wrico pen no. 7T or Leroy pen no. 00B (1 point or 0.013 in.).

    (━━━━━━━━ 1-point rule.)

Narrow line for outlined circles (outside diameter, 0.100 in.), squares (outside length of side, 0.090 in.), or triangles (outside length of side, 0.120 in.) representing the plotted points, and for the full coordinate rulings within the frame or the short stubs indicating the coordinates (¾ point or 0.010 in.). Make circles with a compass (drop bow pen); make squares and triangles with a Leroy pen no. 000B and a cut-out template (such as RapiDesign or a lettering template); make coordinate rulings or scale markers (stubs or ticks) with a ruling pen or a Leroy pen no. 000B.

    (──────── ½-point rule.)

The symbol surrounding each point is inked first. The point at the center of the symbol is not inked; if a pencil dot is present, it is gently erased when the ink is dry. The curves are then drawn to connect or represent the point symbols.

Straight lines are drawn with the aid of a transparent triangle or straightedge, raised above the paper by strips of drafting tape to prevent the ink from running under its edge.

The most difficult line to draw is the wide line used for the curve. Unless special care is taken, the ink will run from this line into the symbols and broaden their lines or even fill them in. The ink of the symbols should be allowed to become thoroughly dry before the curve is drawn. It may be necessary to stop the curve just short of the symbols, and, when the ink is dry, to fill in the gaps by careful retouching with a fine pen (Crow Quill or Hawk Quill). Even a skillful draftsman may have to resort to this expedient when he is drawing a curve with a lettering pen broader than Wrico no. 4 or Leroy no. 3.

A smoothed curve may be drawn in ink with the aid of suitable transparent French curves, elevated from the paper by strips of drafting tape. It is usually necessary to have on hand a number of French curves, and to find sections of these that fit successive segments of the smoothed curve. Care must be taken to have the adjacent segments join smoothly. A flexible curve ruler may be used, instead of French curves, if the curve does not bend sharply.

No curve or coordinate ruling should run through lettering or outlined point symbols. Filled-in symbols falling on the axes may rest in small gaps left in these lines. Where a series of symbols fall along the horizontal axis, it is best to depress the line representing that axis to a negative value.

If full coordinate lines are drawn (fig. 2, 3), they should be spaced from ½ to 1 inch apart in an original having twice the dimensions of the reproduction, so that they will not distract attention from the curve. It is often preferable to omit the ruled grid and to indicate the coordinates by short stubs, or ticks (fig. 4 to 6).

Letters and numbers are made with the aid of Wrico or Leroy lettering guides and pens. Lettering made with guides is neater than free-hand lettering, and it can be duplicated by anyone. The size of the capital letters and numbers should be such that they will be as close as possible to ¼₆ or 60 thousandths of an inch high in the reproduction. Table 8 shows the guides and pens to be used in order to obtain letters and numbers of this size after various degrees of reduction.

The minimum height for clear legibility of lettering in the reproduction is 40 thousandths of an inch (or 1.0 mm). This will be the height of the lower-case letters if the capi-

TABLE 8

Lettering guides to be used for various degrees of reduction to give letters $\frac{1}{16}$ or 60 thousandths of an inch high in the reproduction

FOR REDUCTION OF HEIGHT TO:	DESIRED HEIGHT OF ORIGINAL LETTERING	WRICO		LEROY	
		Guide number*	Pen number	Guide number*	Pen number
	thousandths of an inch				
0.30	200	200	6	200	2
.35	171	175	6	175	2
.40	150	140	7	140	0
.45	133	140	7	140	0
.50	120	120	7	120	0
.55	109	120	7	120	0
.60	100	90	7	100	00
.65	92	90	7	100	00

* Numbers of guides indicate heights of capital letters in thousandths of an inch.

tals and numerals are 60 thousandths, as recommended. It is suitable for lower-case labels running along the curves (fig. 6). But for easier legibility where much lower-case lettering is used in the graph, the capitals should be increased to 70 thousandths of an inch high, so that the lower-case letters will be about 50 thousandths. This size will be obtained if lettering made with guide no. 140 is reduced in reproduction to 0.50 its original height.

Superscripts, subscripts, and exponents are best in appearance if made with a smaller lettering guide, so that they are about 0.8 the height of the letters or figures to which they refer (fig. 4).

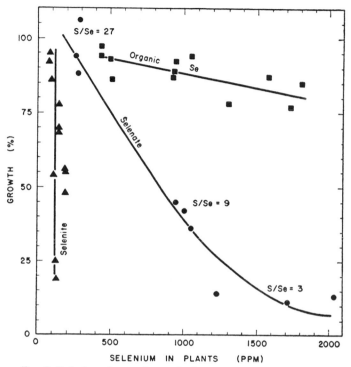

FIG. 6. Relation of growth to selenium content of maize supplied with selenium as selenate, selenite, and organic selenium (derived from extract of *Astragalus*). (*Trelease, S. F., and O. A. Beath. 1949. Selenium.*)

Corrections in the inked graphs are usually made by redrawing after the dried ink has been removed by careful etching with a razor blade and the surface of the paper has been cleaned and smoothed with a soft eraser (not an ink eraser). A graph on opaque paper may be corrected by pasting a strip of paper over the part that needs to be corrected and then redrawing it. Small irregularities in ink lines may be removed with a razor blade or etching knife. Retouching may be done in India ink with a fine pen (Crow Quill or Hawk Quill). After the graph is finished, it should be gently cleaned with a soft eraser, care being taken not to lighten the ink lines.

Some publishers expect the author to provide the numbering and lettering within the graph, but they have the numbers and labels on the coordinate axes set up in type by the printer; the author should place them in pencil outside the axes. Other publishers take care of all numbering and lettering on graphs. In this case, the author is expected to provide each graph with a tracing-paper overlay, fastened with rubber cement or paste to the upper margin of the back of the copy and folded down over the face. Numbers and letters should be written in pencil on the overlay in the exact places where they are to appear on the graph. The style of letters, whether roman or italic (vertical or slant), should be indicated by instructions written on the margins of the graph or tracing-paper overlay.

PHOTOGRAPHS[15]

1. Books. Making good photographs for scientific illustration requires familiarity with the rudiments of photographic technique. Skill is easily gained through practice. Several comprehensive treatises on photography are to be found in most libraries, and a good instruction manual is available, in photographic stores, for each size and type of

[15] Mr. John W. McFarlane, of the Eastman Kodak Co., has helped in the preparation of this section.

camera and even for many individual brands. The Eastman Kodak Co. publishes, and keeps up-to-date by frequent revision, books and other aids giving data on Kodak
products and techniques for their use. These include the
Kodak Reference Handbook, the *Kodak Industrial Handbook* (with sections on "Photography Through the Micro-
scope" and "Making Service Pictures for Industry"), and
the *Kodak Professional Handbook*. They are available
from photographic dealers, either in the bound form or
as separate booklets. Everyone should have a copy of the
Kodak Master Photoguide, which contains in compact
form a great deal of practical information on taking pictures under a variety of conditions outdoors and indoors.

2. Models. It will be useful to study as models the photographs that are published in the various scientific journals.
Critical examination will show good and poor pictures
that serve to emphasize the points discussed here. Perhaps
the best examples of highly effective photography are
those in current advertising material and inexpensive
booklets issued by manufacturers of apparatus, instruments, and photographic materials.

3. Camera. A suitable camera for most scientific subjects is a view camera or a press camera that is focused by
a ground-glass screen, has a double-extension bellows, and
uses sheet films and film packs of medium size—$2\frac{1}{4}$ x $3\frac{1}{4}$,
$3\frac{1}{4}$ x $4\frac{1}{4}$, or 4 x 5 inches. A camera of this type is useful
for close-ups, including copying, and makes negatives that
are large enough for individual handling and for contact
printing of lantern slides. In selecting a lens for scientific
work, preference should usually be given to a symmetrical
or nearly symmetrical anastigmat of moderate speed ($f/5.6$
to $f/7.7$), because of its sharp definition over a wide field
and its superiority for extreme close-ups (examples: Goerz
Dagor, Schneider Symmar, Wollensak Raptar Ia, Kodak

Ektar 8 in. $f/7.7$, and Bausch and Lomb Tessar IIb). The front or rear cell of some of these lenses may be used separately to obtain a focal length approximately twice that of the complete lens. This is advantageous if the subject cannot be approached closely.

Another camera that is likely to be useful to the scientific worker is a 35-mm single-lens reflex, with a ground-glass screen for framing and focusing pictures of either distant or extremely close objects. This camera may be used for most types of photography, and it is especially convenient for making lantern slides of scientific subjects in color. It is a good personal camera for the scientist because it is nearly as convenient as a range-finder camera for ordinary snapshots and has the advantage of introducing no parallax error in making extreme close-ups.

Although cameras of the types just described are technically the best choices for scientific photography, an inexpensive miniature camera, if used with care, is capable of yielding excellent photographs of most subjects. Ready portability, low film cost, and great depth of field are obvious advantages of the miniature camera. By the time the halftone reproductions have been made, it is usually impossible to tell whether a large or small, an expensive or inexpensive, camera has been used in making the original photographs.

Any camera or lens, new or used, should be bought with provision for return of the purchase price if it is found to be unsatisfactory. The most important things to test in a camera are: (a) shutter speeds (calibrated in 10 minutes by a special instrument in a photographic repair shop, or tested by a series of negatives given the same exposure through proper combinations of shutter and diaphragm settings); (b) resolving power of the lens (by photographing charts of the National Bureau of Standards, Circular

533, obtainable from the Government Printing Office); (c) range-finder and focusing scale; and (d) performance under a variety of conditions, including use for close-ups and copying. These tests should be repeated periodically while the camera is in service.

4. Background and Composition. The principal subject of the photograph should be shown as clearly and sharply as possible, and nonessential objects should be subordinated or excluded. The background should be unobtrusive. It should be free from distracting lines or spots and preferably of a uniform tone—white, gray, or black—that contrasts sufficiently with the subject. Light backgrounds are usually more attractive than dark ones. Even a light-toned object may show its form and detail best on a light-gray background. Troublesome shadows, however, can easily be avoided by the use of a black velveteen background. For white or gray backgrounds, pieces of cardboard may be used, or plywood painted with a flat water-base paint.

The objects in the picture should be arranged so as to give a simple and effective composition. Tones of light and shade should be balanced. Point of view and perspective are important. A worm's-eye or a bird's-eye view may sometimes be both effective and pleasing.

5. Illumination. Modeling and texture may be brought out by differential lighting. In photographing scientific specimens, strong left-hand illumination is recommended because we are accustomed to visualize objects as lighted in that way. If photographs of a series of specimens are to be mounted together, illumination in all should come from the same direction. Otherwise elevations may be mistaken for depressions, or vice versa.

The following conventional arrangement of lights is excellent for many scientific subjects. Two lamps of the same strength, such as RFL-2 reflector flood lamps or no. 2 flood lamps in metal reflectors, are used to illuminate the object. The fill-in lamp, which illuminates the shadows, is placed at the same distance from the object as the camera, level with the camera, and about 15 degrees to the right of the camera-object axis. The main lamp, or modeling lamp, is placed ⅔ the distance of the fill-in lamp from the object, about 45 degrees to the left of the camera-object axis, and about 45 degrees above this axis. The background, unless a flat black, may need to be illuminated by a lamp on each side at an angle of about 30 degrees. The same relative positions of the lamps may be used when the camera is either horizontal or vertical. If the shadows appear to be too dark, the fill-in light should be moved closer to the subject, or another light should be added. To obtain a photograph that will yield a halftone reproduction with the desired details in both shadows and highlights, the brightness range of the subject should be kept rather narrow. White or light-colored parts, of course, require less illumination than areas of low reflectance.

Although this simple system of illumination is often satisfactory, it should be modified as much as necessary to suit the particular subject. The important point is to arrange the illumination so that the negative, the print, and finally the halftone reproduction will faithfully record the outline, form, tone values, and details of the original subject.

The arrangement of lamps described above gives the *modeling illumination* that is used for portraits or their equivalents in black-and-white. For *flat illumination,* used to reduce shadows in many black-and-white photographs and in most color photographs, both lamps are placed close to the camera, one to the left of it and the other to the right. A miniature camera may be mounted on a bar (such as the Mayfair Automatic Lite or a motion-picture bar light) that carries the flood lamps. Flash lamps in a suitable reflector are convenient for many types of photographs at subject distances ranging from about 8 inches to 10 feet. For *non-reflecting illumination,* necessary for copying drawings, graphs, printed matter, etc., both lamps are placed level with the camera, one 45 degrees to the left and the other 45 degrees to the right of the camera-object axis. Although this type of lighting is sometimes satisfactory for photographing various types of objects, it may produce harsh shadows.

The new fast films make possible the use of existing light for indoor photographs. If the subject appears satisfactory under such illumination, it can be photographed by a time exposure made on a tripod, or even by a handheld snapshot in a brightly lighted laboratory. An incident-light exposure meter is helpful.

In photographing colored subjects in black-and-white, control of tones requires the use of panchromatic film and a suitable filter. Natural tone contrasts are obtained with a filter that gives full color correction with the particular emulsion and light source used. When desired, exaggerated contrasts may be secured with a properly chosen filter. A filter lightens (in the print) the rendering of objects of its own color as compared with objects of other colors. The effect of strong filters—red, orange, green, and blue—can be seen through the filter beforehand. The viewing filters in the *Kodak Master Photoguide* simplify selecting the appropriate filter.

6. Close-ups. Extreme close-ups of small objects are often needed in illustrating scientific articles. Information on this type of photography is contained in the Kodak Industrial Data Book entitled *Making Service Pictures for Industry.*

Black-and-white close-ups are easily made with a sheet-film camera having a double-extension bellows and ground-glass focusing. If necessary, a four-fold enlargement of any part of the negative may usually be made in printing, without perceptible loss of detail. A 35-mm single-lens reflex is convenient for close-ups in black-and-white or color. When used with positive supplementary lenses and extension tubes or a bellows, this camera provides accurate ground-glass framing and focusing of photographs up to natural size.

Even a simple camera can be used for close-ups if a positive supplementary lens is added to the camera lens and a suitable attachment is used to correct parallax error and permit accurate focusing and determination of field size. Various close-up devices are available for almost every type of camera. One of these is a focal frame attached to the camera base. This frame surrounds the field photographed and assures correct focus. It can easily be built by a person who is handy with tools. (See articles on *Portra Lenses and a Technique for Extreme Close-ups* and *Flower Pictures in Color,* available on request to the Eastman Kodak Co.) An adjustable focal frame can be purchased under the name Cal-Cam Focus Guide. The Eastman Kodak Co. supplies focal frames for the Retina IIIc camera, and provides a close-up kit that serves as a focal frame of 4-inch width for several miniature cameras. Since the camera is used with the normal lens-to-film distance, ordinary exposure guides apply. This technique is useful for both outdoor and indoor work with a 35-mm camera that lacks ground-glass focusing, and it is adaptable to a wide range of subjects—flowers, insects, small animals, apparatus, and medical subjects, such as skin, teeth, eyes, and pathological specimens.

Low-power photomicrography, giving magnifications from 1 to about 25 diameters, requires a long bellows draw. A series of special lenses is available for such photographs—Bausch and Lomb Micro-Tessar, Goerz Dagor, and Schneider Symmar—with focal lengths ranging upward from 16 mm. An ordinary camera lens or enlarger lens of the unsymmetrical type is satisfactory if it is reversed, so that the normally front, engraved cell faces the film (the long conjugate side in this case). This is practicable if the camera has a removable lens-board. A lens with a focal length of 50 mm is especially suitable for magnifications up to 5 or 10 diameters. The second formula given may be used to calculate the magnification obtainable with any specified bellows length and focal length. For magnifications greater than unity, it is best to use a vertical camera on a sturdy rod support, such as a photomicrographic camera or the Kodak Flurolite Enlarger with a camera-back adapter.

When the bellows of the camera is considerably extended, for subjects closer than about 8 times the focal length of the lens, the exposure must be multiplied by the correct factor. This may be ascertained with the aid of the Effective Aperture Computer in the *Kodak Master Photoguide,* or it may easily be calculated by means of the first formula given below.

Exposure factor $= $ (Lens-to-film distance)^2 $\div$ (Focal length)^2
Magnification
$\quad = $ (Lens-to-film distance $-$ Focal length) $\div$ (Focal length)
Lens-to-subject distance
$\quad = $ (Focal length $\div$ Magnification) $+$ (Focal length)

7. Trial Exposures. In all close-up photography, correct exposure can best be attained by a series of tests. A good procedure is to make an estimate of the correct exposure, preferably with the aid of a meter, and then to try four exposures—$\frac{1}{3}$, $\frac{2}{3}$, $\frac{4}{3}$, and $\frac{8}{3}$ times this estimate. The films should be developed while the camera is in position. This method rarely fails to produce at least one satisfactory negative. But, of course, if the range has been missed, a new series must be used. After the correct exposure has been determined, it may be used as the basis for other photographs of the same type of subject if due allowance is made for any change in bellows length.

For ordinary black-and-white photographs outdoors a good series is $\frac{1}{4}$, 1, and 4 times, or for color photographs $\frac{1}{2}$, 1, and 2 times, the exposure estimated with the aid of a meter or guide. These exposures are obtained by changing the diaphragm setting by steps of 2 stops and of 1 stop, respectively. The routine use of a range of exposures gives best assurance of obtaining a good picture. Err, if you must, slightly on the side of underexposure of a color photograph, or of overexposure of a black-and-white.

Although an exposure meter, used properly, is a valuable aid under unusual light conditions, it is not essential in photographing normal subjects in bright or hazy sunlight, or even in floodlight. In such photography, correct exposure may be obtained by following the *Kodak Master Photoguide* or the instructions packed with the film. If a significantly different exposure is indicated by a meter, one should suspect the accuracy of the meter or the method of using it. It is well to follow carefully the instructions that come with the meter, and to consult *Exposure Indexes and How to Use Them,* published by the Eastman Kodak Co.

8. Copy for Reproduction. Photographs are reproduced as halftones, in which the picture is broken up into minute dots. The photograph for copy should be as clear and sharp as possible. It may be considerably larger than the reproduction, or the same size, but should never be smaller. If a print needs to be retouched, it should be of large size—5 x 7 or 8 x 10 inches. But if retouching is unnecessary, a print of the same size as the intended reproduction has the advantage of allowing the quality of the reproduction to be better predicted.

Loss of fine detail is inevitable in reproduction by the halftone process. To compensate for this loss, the scale of magnification must be larger in a halftone than would be satisfactory in a glossy photographic print. For example, if one-fifth natural size is a sufficiently great magnification to show the details of an object well in a glossy print, double this magnification, or two-fifths natural size, may be required in a halftone to make the details equally clear. Disappointment is certain to follow if the opposite procedure is adopted—for example, if a print that barely shows essential details is reduced to half-scale in reproduction. Use close-ups of essential features; and in making the print, enlarge sufficiently and then crop away all extraneous parts of the photograph.

Glossy white ferrotyped paper is best for prints intended for reproduction. A paper with a pebbled or rough surface, or a cream color, should never be used.

9. Contrast of Print. The print for reproduction should preserve as closely as possible the details that exist in the original subject. It is desirable to have the print show a moderately wide range of tone values, with detail in both highlights and shadows. A good procedure is to make a series of prints on several contrast grades of paper, and then to select the print that seems best.

The usual method of selecting the proper contrast grade of paper for a normal print intended for viewing is to make a test print that gives the desired highlight gradations. Then: (*a*) if the shadows are also correct, the paper is of the right contrast grade; (*b*) if the shadows are blocked by overexposure, use a paper of less contrast; (*c*) if the shadows are not dark enough, use a paper of more contrast.

Special requirements for halftone reproduction necessitate a modification of the usual procedure in print making. A print for reproduction should not make use of the full range of tone values from clear white to jet black, because the rendering of detail at both ends of the tone scale is unavoidably degraded in the print. To prevent loss of highlight and shadow gradation in the halftone reproduction, a print of slightly softened quality is therefore recommended.

The best method of making such a print is to use the same contrast grade of paper that has been found by trial to give a normal print, but to adopt the following procedure: Adjust the printing exposure so that the lightest important highlight detail is a very light gray, not a clear white. After obtaining this exposure by trial, make the final print in a manner that holds back the extreme shadow regions by sufficient dodging to prevent the darkest important shadow detail from becoming a full black. When dodging is impracticable, use a paper that is one contrast-grade lower than the paper that gives a normal print for viewing, and adjust the exposure so that the highlights are light gray and the shadows are less than jet black.

The use of a slightly softened print for reproduction does not result in a reproduction that is too soft. In the hands of a good photoengraver, the tone scale is expanded so that a much better reproduction will be obtained than could result from a full-scale normal print.

10. Blemishes and Retouching. Blemishes in a print

will be conspicuous in the reproduction. Imperfections of this sort include blurred images, muddy highlights, fog, spots, scratches, dents, cracks, and stains. A defective print should preferably be replaced by one that is perfect. But very small blemishes, such a white spots, can be concealed with a soft lead pencil or water-color paint. Numbers, symbols, letters, arrows, etc., may be added in India ink. A glossy print can be prepared so that it will accept retouching by sifting Fuller's earth or talc on the print and then rubbing this very lightly over the surface with a dry tuft of cotton. A small area can be prepared by rubbing carefully with a soft eraser. Skillful retouching by an artist may remove defects in a print, but the work is expensive and often unsatisfactory.

If it is necessary to write on the back of a photograph, lay the print on a smooth, hard surface (such as a sheet of glass) and write very lightly with a soft pencil. Unless these precautions are taken, the writing will show in relief on the face of the print. A better method is to write on a separate strip of paper and attach this with dry mounting tissue to the margin on the back of the print.

11. Mounting. The print should be mounted on a piece of smooth flat cardboard (Bainbridge board no. 80), large enough to leave a margin of at least an inch around the print. Dry mounting tissue is best for this purpose. Rubber cement is convenient for temporary mounting, but after some time it is likely to produce a discoloration of the photograph that will show in the halftone. Paste, glue, and mucilage should be avoided, because they wrinkle or stain prints. It is usually best not to trim the print, before mounting, to the exact size desired. Cropping lines should be drawn in pencil on the margin of the cardboard mount, extending to the print but not across its face. The photoengraver will include only the indicated part of the print and will square it. The mounted photograph should be covered with a protective flap of brown paper attached with paste to the upper edge of the rear surface of the cardboard and folded down over the face of the print.

12. Arrangement in Groups. When a number of separate photographs are to be arranged together for halftone reproduction in a single plate, they should be matched for uniformity in density and contrast. If the group consists of some light and some dark prints or includes prints that differ in contrast, these differences may be accentuated in the reproduction.

The selected prints should be trimmed and mounted with special care. If the publisher prefers to do the mounting in groups, the prints should be supplied untrimmed and unmounted. Unless the mounting is done properly, the background will have to be routed out, thus doubling or trebling the cost of reproduction. This extra cost is usually chargeable to the author. If possible, the photographs should be fitted together perfectly, so as to cover completely the cardboard (Bainbridge board no. 80) on which they are mounted. The photoengraver will cut white lines where they join. Rubber cement is most convenient for such mounting because it allows the position of each print to be adjusted accurately.

Figure numbers and explanatory letters may be put directly on the photographs in black ink or white ink with the aid of a Wrico or a Leroy lettering guide (see table 8). Printed characters may be attached with rubber cement. Light-faced characters should be used, and all the slips of paper that bear them should be the same size, trimmed square, and set straight in mounting, since they are always visible in the halftone.

LANTERN SLIDES[16]

1. Visibility. Lantern slides prepared to illustrate a scientific or technical lecture should contain much less material than can be adequately shown in a printed illustration or table. A slide should be so clear and simple that persons with normal eyesight sitting in the back of the room can see it easily and grasp its meaning completely. It is just as important to have slides that are visible to everyone as it is to speak so that the whole audience can hear. Slides should show essential features very clearly, and should be as free as possible of nonessentials that distract the eye.

For easy legibility, the minimum height of the lettering on a well-illuminated screen should be $\frac{1}{300}$ the maximum viewing distance. This means 1-inch lettering for a distance of 25 feet, 2-inch lettering for a distance of 50 feet, etc. Lettering of this height will be obtained on the screen if the slides have the lettering recommended below and if the projection conditions are usual (maximum viewing distance 6 times the screen height or 4 times the screen width).

2. Rule for Letter Height. For good legibility, the minimum letter height should be $\frac{1}{50}$ the total screen-image height. The letter height on the slide should therefore be $\frac{1}{50}$ the shorter dimension of the standard slide-mask opening. When preparing an original for a slide within a rectangle having the proportions of the slide-mask opening, the lettering should be made $\frac{1}{50}$ the shorter dimension of the rectangle.

[16] Mr. John W. McFarlane, of the Eastman Kodak Co., has helped in the preparation of this section.

Additional information on making slides is given by the American Standards Association (1932), Bonnell (1949), and the Eastman Kodak Co. (1955).

Slides used to illustrate a popular lecture should have especially clear lettering. For these, the smallest size may well be $\frac{1}{40}$ the normal image height, and headings $\frac{1}{30}$.

3. Tables. It would be a mistake to attempt to show, in a single slide, a full-page comprehensive table of data. The table would be too complex to be readily understood, and it would be illegible in the back of the room. A legible slide cannot be made from an ordinary printed table that is larger than about $3\frac{1}{4}$ x $4\frac{1}{2}$ inches. Printed tables, even if not too large, usually have headings that contain too many words, and lines of figures that are too closely spaced. Best results are obtained by preparing simplified typewritten tables as originals for slides.

To insure lettering of adequate size, the original table should be typewritten, double-spaced, in pica characters (10 to the inch) so that it fits within a 5 x 7-inch rectangle, with small margins all around. The lower-case letters are $\frac{1}{5}$ of 5 inches in height. (If elite characters are used, the rectangle should be only $4\frac{1}{4}$ x 6 inches.) For a $3\frac{1}{4}$ x 4-inch slide, the shorter dimension of the rectangle must be vertical, but for a 2 x 2-inch slide either dimension may be vertical. The title should be limited to one line of about seven words, and the data to three or four columns of about ten items in each.

The typewritten original may be made by using a new, well-inked cloth ribbon and carefully cleaned type. Clearest lettering is obtained by typing through a carbon ribbon on an electric typewriter. If the table is typewritten on a sheet of ordinary bond paper, the sheet should be backed with a sheet of carbon paper that is reversed so that it will print on the rear surface. To correct an error, it is best to retype the whole table. A less satisfactory method is to type the correction on a slip of paper and cement this over the erased original.

4. Graphs. Graphs to be used as slides should be very simple, with broad lines and large numbers and letters. Numerous, crowded curves cannot be deciphered by the audience.

A convenient way to prepare a slide that has lettering of suitable size is to compose the original graph—including scale labels and clear margins—within a 6 x 8-inch rectangle for a $3\frac{1}{4}$ x 4-inch slide, or a 6 x $8\frac{3}{4}$-inch rectangle for a 2 x 2-inch slide, and to use a lettering guide that gives capital letters and numbers 0.120 inch high. The letters are $\frac{1}{50}$ the height of the rectangle (6 $\times$ $\frac{1}{50}$ = 0.120). A title of not more than five words may be placed in 0.175-inch letters above or below the graph, or in a balanced position within it. The graph is most easily drawn, first in pencil and finally in India ink, on coordinate paper ruled with nonphotographing light-blue lines. Since a $3\frac{1}{4}$ x 4-inch slide must be used in the projector with its $3\frac{1}{4}$-inch edge vertical, the original graph should be no higher than the shorter dimension (6 inches) of the rectangle if the slide is to be fitted by the standard mask ($2\frac{1}{4}$ inches high). But for a 2 x 2-inch slide, the original may be prepared with either dimension vertical. It is recommended that the curves be drawn with a Leroy no. 3 pen or a Wrico no. 4 pen, and that the other specifications given on page 144 for line widths also be used.

In making the slides, the originals should be reduced no more than necessary to have them fit the standard slide-mask sizes, which are $2\frac{1}{4}$ inches high by 3 inches wide for $3\frac{1}{4}$ x 4-inch slides, and $2\frac{9}{32}$ x $1\frac{5}{16}$ inches or $1\frac{1}{32}$ x $1\frac{1}{2}$ inches for 2 x 2-inch slides.

Originals that have been prepared for publication, or their printed reproductions, are often unsuitable for slides. They may be too complex or may have lettering that will be too small after they are reduced to fit the maximum usable area of the slide. This maximum area is $2\frac{3}{4}$ inches high and about $3\frac{1}{4}$ inches wide on a $3\frac{1}{4}$ x 4-inch slide or about $1\frac{1}{2}$ x $1\frac{1}{2}$ inches on a 2 x 2-inch slide. Each original should be checked critically to determine whether it will make a good slide. Height of lettering can be estimated with sufficient accuracy by using a hand lens and a transparent ruler, preferably marked in tenths of an inch.

In many cases it will be found that new originals must be prepared expressly for slides. This may usually be done, without replotting the data, by tracing the graph in pencil (HB or F grade), either from the original or from a suitably reduced or enlarged photographic copy (e.g., Photostat copy), on high-grade tracing paper, simplifying it if necessary, and inking it in with proper lines and lettering.

5. Complex Objects. When illustrating complex three-dimensional objects by lantern slides, it is best to give first a comprehensive view for orientation, and then to show a series of close-ups in which the individual features are presented on as large a scale as possible.

6. Presenting the Paper. Most papers presented at meetings of the professional societies are based upon articles prepared for publication in scientific or technical journals. Those attending the meeting want to see the speaker, get a first-hand account of his work, and have an opportunity to ask him questions at the end of the talk. They are interested in a paper that brings out something new—a new idea, relation, conclusion, or interpretation, supported by evidence in the form of original data.

The draft that has been prepared for publication is almost always unsuitable for oral presentation with lantern-slide illustrations. It usually contains too much material and too many details, and its organization is usually too complex.

A simplified version of the paper should be prepared. This should be so direct and clear that it will be easily understood by the audience. Remember that the audience is almost certain to include many persons who are not specialists in your immediate field. Make a detailed outline, with marginal notes of the slides that are to be shown. Then, if you wish, write out the paper. But preferably speak from a brief topical outline, which may be consulted if necessary. A good way to keep from being nervous is to be well prepared.

At the beginning, spend a few minutes giving briefly the general information that the audience needs for understanding the subject. Then state clearly the object of your work, or even the principal conclusion reached, so that the audience can appreciate and weigh the evidence that you will present. Select a few of your most important points and present them so vividly that they will be easily understood by the audience. The interpretation of the results is the most interesting and important part of your paper. Indicate, by means of simplified tables and graphs, the nature of the evidence, but do not try to give all the details. Tell what each slide shows, and what conclusions may be drawn from it. As a rule, no slide should be on the screen for less than half a minute or for more than three minutes.

Speak clearly in your natural conversational tone, and loudly enough to be easily heard by those in the back of the room. Face the audience, not the screen. It is easier and more effective to use simple, direct, informal language in an oral report, rather than the more complicated English usually found in a printed article. But this does not mean being careless in the choice of language.

If you feel that you must read your paper, instead of depending on a brief outline, practice reading in your natural manner of speaking. Avoid a monotonous delivery and read slowly. It is quite possible to learn to read from script, as have many radio and television speakers, in a manner that resembles ordinary speech.

Prepare your paper so that it can be kept strictly within the allotted time. To do this, several rehearsals, with revisions, are likely to be required.

Welcome discussion and criticism of your paper by the audience. Valuable new ideas are likely to come from such discussion.

PREPARING ILLUSTRATION COPY

1. Identification. For purposes of identification, the figure number, the author's name and address, and the title of the article should be written on the margin or back of each piece of illustration copy, or on a piece of paper attached securely with paste to the lower margin of the copy. The "top" of the illustration should be indicated if there is any possibility of misunderstanding.

2. Directions for Reduction. Clear directions for reduction should also be written on the margin or back. (See section on "Correct Proportions," page 123.) In giving directions, it is best to specify the final width or height. (For example: "Reduce width to 3 inches" or "reduce height to 6 inches.") In designating fractional reduction, it is better to say "reduce width to ¼" than "reduce ¾" or "¾ off."

3. Photographic Copies. If the illustrations are larger than 8½ by 11 inches, duplicate photographic prints or Photostat copies of smaller size should accompany the manuscript, to facilitate sending the article to reviewers.

Retain a good photographic copy of each illustration, for use if the original is lost in the mail.

The original is best for making a photoengraving. A very good photographic copy may be used. But a copy that is out of square (not rectangular) or faint is unsuitable.

4. Legends. The legends, or titles, of plates and figures should be self-explanatory. They should be typewritten *double-spaced* in numerical order upon one or more sheets of paper, which are placed at the end of the manuscript, after the literature cited. Always supply a short title for the illustration. Any descriptive matter may follow the title, in the form of paragraphs.

The legend of each text figure is printed below the figure. A short title appears below each plate, and complete descriptions of all plates are usually given in a separate section of the paper, after the literature cited and before the plates.

5. Place of Insertion. The place of insertion of each text figure must be marked in the manuscript and in the galley proof. (For example, write in the margin: "Insert figure 2.")

6. Numbering. Text figures should be numbered from 1 up in each article. Plates should be numbered 1, 2, 3, etc., in each article; and figures in plates should be numbered from 1 up, beginning a new series either in each plate or in each article. Some journals number all figures (in text and plates) consecutively from 1 up in each article; this simplifies text reference to the figures.

7. Reference in Text. In the text, the figures and plates should be referred to by number; the words *figure* and *plate* should not be capitalized. (For example: Examination of figure 5 of plate 3 shows that . . . . . .) If the reference is made parenthetically, the words *figure* and *plate* should be abbreviated, using the forms "fig." and "pl." for both singular and plural. [For example: The data of table 7 are shown as graphs in figure 4, in which the method of plotting is the same as for series A (fig. 1).]

Be sure to check all text references to illustrations after the manuscript has been completed.

SHIPPING ILLUSTRATIONS

Photographic prints or drawings intended for halftone reproduction are likely to be damaged when sent by mail or express unless they are well protected, especially at the corners. The following method of wrapping usually gives good protection: Place the prints between sheets of thin cardboard, cut to a size slightly larger than the prints. If the prints are mounted, cover them with a sheet of thin cardboard of the same size as that on which they are mounted. Bind the cardboard sheets together on all four sides with short strips of cellulose tape. Anchor this packet securely, with more strips of cellulose tape, to a piece of stout corrugated board about two inches larger all around than the original packet. This will keep the packet from slipping to an edge or corner. Place another piece of corrugated board of the same size on top (preferably one with the corrugations running at right angles to those of the other), and bind the two firmly together with strips of cellulose tape. Finally, wrap in heavy paper, and seal all loose edges with gummed tape or tie securely with string. Send by first-class mail, registered, or by express, whichever costs less.

Chapter 6

PREPUBLICATION REVIEW

1. Purposes. Many scientific and technical journals have adopted the plan of having every paper that is submitted for publication read and critized by two competent reviewers selected by the editor or the editorial committee. The purposes of this procedure are (1) to improve the quality of the papers that are printed in the journal and (2) to avoid the acceptance of material more rapidly than it can be published with the funds available. Acceptance of material is limited by promoting condensation of text and tabular material and elimination of unessential illustrations, as well as by declining the papers that make the least distinct contributions to the particular field of science. Since all papers are sent to reviewers, this procedure implies no reflection on the merits of the papers. Prepublication review represents an editorial service that the authors appreciate in the majority of cases.

2. Work of Reviewers. Each reviewer is asked to give his general opinion regarding the suitability of the paper for publication in the journal, and to make specific suggestions regarding possible errors, lack of clearness, parts that may be condensed, omitted, or improved in form and arrangement, etc. The reviewer may be asked the folowing questions:

(*a*) Would you grade the paper A, B, C, D, or E, on the basis of its relative merit as a scientific contribution—if C represents the average rank of papers in recent volumes of the journal? (*b*) Has the material been published previously? (*c*) Has the work been carried far enough to warrant publication? (*d*) Is there some other journal for

169

which the paper would be more suitable? (*e*) Are the conclusions logical and are they based on accurate and sufficient data? (*f*) Is the arrangement logical? If not, suggest improvements. (*g*) Which, if any, of the main ideas are not developed with sufficient emphasis? (*h*) What parts may be condensed or omitted? (*i*) Have you found any errors in the paper? (*j*) Is there lack of clearness? If so, where? (*k*) Where does the literary form need to be improved? (*l*) What improvements, if any, do you regard as necessary in the illustrations? (*m*) Which, if any, of the illustrations could be omitted?

3. Author's Revision. If the reviews indicate that the article would be acceptable but needs revision, it is returned to the author with the comments of the reviewers (quoted anonymously so that the matter of personalities will not enter) and a note that asks the author to study the paper again with regard to revision in accordance with the reviewers' suggestions. The author is told that if he does not consider it desirable to adopt certain of the recommendations of the reviewers, the reasons for his preference should be explained. The author must assume full responsibility for the content of the paper. When the paper has been revised by the author, it is returned to the editor.

4. Judgment of a Third Reviewer. Advice of a third reviewer may be asked by the editors if the two reviewers disagree as to whether the paper would be acceptable after revision, or if the author is unwilling to revise the paper in accordance with the reviewers' recommendations. The opinions of reviewers are advisory, and final responsibility for the selection of papers rests with the editor.

5. Publication after Acceptance. After its acceptance the paper is published in its proper turn, according to the original date of receipt, unless revisions necessitate unavoidable delay.

6. Rejection of Manuscripts. The editor may decide, after seeing the reviewers' comments and reading the paper himself, that the paper could not be accepted even if it were revised. In this case he returns it to the author with a brief note of regret, containing suggestions, if possible, regarding suitable journals to which the article might be submitted. The editor obviously must not accept material more rapidly than it can be published with the funds available. With the aid of the reviewers and the editorial committee, he selects the papers that seem to be best, and he is compelled to reject the others.

Chapter 7

PROOFREADING

GALLEY PROOFS

Galley proofs, on sheets about 20 cm wide and 60 cm long, are submitted to the author, together with the manuscript. The author is expected to correct the proofs; he should see that the proofs agree with the manuscript, and should correct all genuine errors. The proofs should be returned to the editor as soon as possible.

All corrections must be made by means of proofreader's marks in the margins of the proof sheets. Corrections should be made clearly and neatly, using a *red pencil for printer's errors* and a *black pencil for changes from copy.* They should be made horizontally on the page, and opposite the printed lines in which the errors occur.

METHOD

1. Two Persons. If possible, have another person slowly read aloud from the manuscript while you follow the galley proofs and make the necessary corrections and changes. The one who reads aloud should call your attention to every paragraph, mark of punctuation, capitalized word, italicized figure or word, boldfaced figure or word, etc., and should spell out all unusual names or technical terms. If you cannot secure the services of another person in this work, it will be necessary for you to compare carefully the galley proofs with the manuscript, line by line or sentence by sentence.

2. Two Readings. Always read the proofs *twice,* at least.

172

MISCELLANEOUS SUGGESTIONS

1. Expense of Alterations. Author's alterations, or changes from the original copy sent to the printer, are very expensive, and some journals charge them to the author. Only real errors should be corrected in the proofs. Each line reset may cost the author twenty-five or fifty cents. A minor insertion or deletion may require resetting all the lines that follow in the paragraph, unless the number of characters in the altered proof is kept the same as that in the original proof, by an equivalent deletion or insertion (see section on "Estimating the Length of the Printed Article," page 62).

2. Special Attention. Give particular attention to tables, figures, names, quotations, and citations. Check text references to illustrations. Assume that errors are present; find and correct them.

3. Questions. Be sure to answer questions, or queries, made by the printer.

4. Instructions to Printer. If you do not know how to indicate a correction, simply draw a horizontal line through the word that needs to be changed and then write clear instructions in the margin, enclosing the instructions in a circle.

5. Omissions. Watch for words or lines that may have been omitted.

6. Reading for Meaning. After you have read the proofs *twice,* as suggested above, it is well to read them a *third* time, paying particular attention to the sense, or meaning, of the statements. You will not be permitted to make revisions; but genuine errors must be corrected, of course, whenever they are discovered.

7. Tables. Check to make sure that the tables have been properly distributed, or that their positions have been correctly marked in the margin of the proof.

Cancellation

Delete, or take out, character or ~~the~~ word marked.

Insertion

Insert ^ word, letter, or punctuation mark written in the margin.

Spacing

Insert space between ^ words, letters, or lines.

Close up, or take out the space.

Close up, but leave some space.

Position

Turn a reversed letter.

[Carry farther to the left.

Carry farther to the right.

Move down a letter, character, or word.

Move up a letter, character, or word.

Indent one em.

Straighten a crooked line.

| Straighten lateral margin of printing.

Transpose of order words letters.

Correct uneven spacing.

Paragraphing

Make a new paragraph.

No paragraph.

Miscellaneous

Push down a space or quadrat that **prints**.

Question to author. Is this right?

Allow to ~~stand~~ as it is.

Kinds of type

l. c. Put in lower case.

caps Put in ~~capitals.~~

U Use a capital.

s.c. Put in small capitals.

w.t. Put in small capitals.

rom. Put in ~~Roman.~~

ital. Put in ~~italic.~~

b.f. Put in ~~bold face.~~

Put in bold-face.

w. f. Wrong font (wrong size or style).

Superscript 4.

Superscript 1.

Subscript 2.

X Type is broken or imperfect.

Punctuation

Period.

Comma.

Semicolon.

Colon.

Apostrophe.

Quotation marks.

Hyphen (-).

One-en dash (–).

One-em dash (—).

Two-em dash (——).

Parentheses.

Brackets.

8. Illustrations. The approximate place for inserting every illustration should be clearly marked in the margin of the proof. (For example, write in the margin: Insert figure 1.) Check the magnifications of drawings and photographs in the photoengraver's proofs, and correct the legends if necessary.

Examine the photoengraver's proofs for blemishes, such as broken lines or letters and extraneous dots or lines. If a correction is needed, direct attention to it by means of a red arrow or circle and write your comments in the margin, before returning the proof to the editor. Do not send the original illustration unless requested to do so.

9. Headings. Look through the proofs for the purpose of correcting errors in all headings.

PAGE PROOFS

After the corrected galley proofs have been received by the editor and have been read and marked by him, they are returned to the printer. The corrections marked on the galley proofs are made, and then the type is divided into pages of the required length. Page proofs are sent to the editor, who compares them with the galley proofs to see that all the corrections have been made and that new errors have not been introduced into the lines that have been reset. A correction is made in Linotype by resetting the whole line; it is made in Monotype by handling the individual characters. The editor then reads the page proofs critically, searching for inconsistencies or errors. The page proofs are returned to the printer, who makes the necessary changes and begins the actual press work.

BIBLIOGRAPHY

Adams, J. K. 1955. Basic statistical concepts. 304 p. New York: McGraw-Hill Book Co.

American Standards Association. 1932. Engineering and scientific charts for lantern slides (Z15.1—1932; reaffirmed 1947). New York: American Society of Mechanical Engineers.

American Standards Association. 1943. Engineering and scientific graphs for publications (Z15.3—1943; reaffirmed 1947). New York: American Society of Mechanical Engineers.

Arader, H. F. 1955. A system for testing and increasing the intelligibility of technical reports. Science 121: 537–539.

Ball, J., and C. B. Williams. 1955. Report writing. 407 p. New York: Ronald Press Co.

Barzun, J., and H. F. Graff. 1957. The modern researcher. 386 p. New York: Harcourt, Brace and Co.

Beveridge, W. I. B. 1951. The art of scientific investigation. 171 p. New York: W. W. Norton and Co.

Bonnell, L. S. 1949. Preparation of effective lantern slides. Chem. and Engineer. News 27: 2600–2606.

Brennecke, E., Jr., and D. L. Clark. 1942. Magazine article writing. 486 p. New York: The Macmillan Co.

Campbell, W. G. 1954. Form and style in thesis writing. 114 p. Boston: Houghton Mifflin Co.

Candelaria, N. 1955. Security and the editor. Science 121: 528–530.

Chicago, University Press. 1952. A manual of style. 11th ed., 521 p. Chicago: University of Chicago Press.

Christman, R. C. 1954. Illustrations for scientific publications. Science 119: 534–536.

Cleland, R. E. 1955. The use of material. Science 121: 519–523.

Cochran, W. G., and G. M. Cox. 1950. Experimental designs. 454 p. New York: John Wiley and Sons.

Cope, A. C., E. L. Haenisch, L. P. Hammett, W. S. Johnson, E. O. Wiig, W. G. Young, and J. H. Howard. 1957. Doctoral training in chemistry. Chem. and Engineer. News 35: 56–60, 65–67.

Crouch, W. G., and R. L. Zetler. 1954. A guide to technical writing. 2nd ed., 441 p. New York: The Ronald Press Co.

Croxton, F. E. 1953. Elementary statistics with applications in medicine. 376 p. New York: Prentice-Hall.

Croxton, F. E., and D. J. Cowden. 1939. Applied general statistics. 944 p. New York: Prentice-Hall.

Dixon, W. J., and F. J. Massey, Jr. 1951. Introduction to statistical analysis. 370 p. New York: McGraw-Hill Book Co.

Eastman Kodak Co. 1955. Photographic production of slides and film strips. 52 p. Rochester, N. Y.

Elder, J. D. 1954. Jargon—good and bad. Science 119: 536–538.

Emberger, M. R., and M. R. Hall. 1955. Scientific writing. 468 p. New York: Harcourt, Brace and Co.

Federer, W. T. 1955. Experimental design. 544 p. New York: The Macmillan Co.

Fishbein, M. 1948. Medical writing. 2nd ed., 292 p. Philadelphia: Blakiston Co.

Fisher, R. A. 1951. The design of experiments. 6th ed., 244 p. New York: Hafner Publishing Co.

Fisher, R. A. 1954. Statistical methods for research workers. 12th ed., 356 p. Edinburgh: Oliver and Boyd.

Gaum, C. G., H. F. Graves, and L. S. S. Hoffman. 1950. Report writing. 3rd ed., 384 p. New York: Prentice-Hall.

Geist, R. J., and R. A. Summers. 1951. Current English composition. 582 p. New York: Rinehart and Co.

Gill, R. S. 1949. The author-publisher-printer complex. 2nd ed., 144 p. Baltimore: Williams and Wilkins Co.

Goulden, C. G. 1952. Methods of statistical analysis. 2nd ed., 467 p. New York: John Wiley and Sons.

Hald, A. 1952. Statistical theory with engineering applications. 783 p. New York: John Wiley and Sons.

Hale, D. H. 1955. Use of the technical report in military planning. Science 121: 532–534.

Hall, P. G. 1954. Introduction to mathematical statistics. 2nd ed., 331 p. New York: John Wiley and Sons.

Higgins Ink Co. 1948. Techniques. 5th ed., 48 p. Brooklyn, N. Y.

Higgins Ink Co. 1953. Technical illustration. 90 p. Brooklyn, N. Y.

Hildebrand, J. H. 1957. Science in the making. 116 p. New York: Columbia University Press.

Kapp, R. O. 1948. The presentation of technical information. 147 p. London: Constable and Co.

Kierzek, J. M. 1954. The Macmillan handbook of English. 3rd ed., 579 p. New York: The Macmillan Co.

Lee, M. O. 1954. Problems in financial management of scientific journals. Science 119: 530–532.

Lohr, L. R. 1932. Magazine publishing. 328 p. Baltimore: Williams and Wilkins Co.

McAllister, D. T. 1955. Is there accepted scientific jargon? Science 121: 530–532.

McCartney, E. S. 1953. Recurrent maladies in scholarly writing. 141 p. Ann Arbor, Mich.: University of Michigan Press.

McCartney, E. S. 1954. Does writing make an exact man? Science 119: 525–528.

Marschalk, H. E. 1955. Technical manuals: their increasing importance to industry and defense. Science 121: 539–540.

Miller, R. C. 1955. The care and training of authors. Science 121: 526–528.

Mills, G. H., and J. A. Walter. 1954. Technical writing. 463 p. New York: Rinehart and Co.

Nelson, J. R. 1952. Writing the technical report. 3rd ed., 356 p. New York: McGraw-Hill Book Co.

Noyes, W. A., Jr. 1954. Probable trends in scientific publications as viewed from the editor's office. Science 119: 529–530.

Oxford Dictionary of Current English, Concise. 1951. Rev. by E. McIntosh. 4th ed., 1536 p. Oxford: Clarendon Press.

Perrin, P. G. 1950. Writer's guide and index to English. Rev. ed., 833 p. Chicago: Scott, Foresman and Co.

Price, W. C. 1954. Preparing manuscripts for Phytopathology. Phytopathol. 44: 667–674.

Ramsay, A. M. 1927. Clinical research by family doctor. Brit. Med. Jour. 2: 1215–1217.

Ridgway, J. L. 1938. Scientific illustration. 173 p. Stanford University, Calif.: Stanford University Press.

Riker, A. J. 1952. Literature citations: how biologists like them. Amer. Inst. Biol. Sci. Bull. 2: 18–19.

Roget, P. M. 1931. Thesaurus of the English language in dictionary form. Rev. by C. O. S. Mawson. 600 p. Garden City, N.Y.: Garden City Publishing Co.

Sanderson, R. T. 1955. Do ye unto others. Science 121: 569–570.

Shillaber, C. P. 1944. Photomicrography in theory and practice. 773 p. New York: John Wiley and Sons.

Skillin, M. E., and R. M. Gay. 1948. Words into type. 585 p. New York: Appleton-Century-Crofts.

Soule, R. 1938. A dictionary of English synonyms. Rev. and enl. by A. D. Sheffield. 614 p. Boston: Little, Brown and Co.

Staniland, L. N. 1953. The principles of line illustration. 212 p. Cambridge, Mass: Harvard University Press.

Struck, H. R. 1954. Recommended diet for padded writing. Science 119: 522–524.

Summey, G., Jr. 1949. American punctuation. 182 p. New York: Ronald Press Co.

Thomas, J. D. 1949. Composition for technical students. 460 p. New York: Charles Scribner's Sons.

Townley, K. A. 1955. Clarity in geological writing. Science 121: 535–537.

Tulloch, G. S. 1954. Problems of the editor of a small journal. Science 119: 532–534.

Tumbleson, R., and H. L. Brownson. 1954. Survey of operations and finances of scientific journals. Science 119: 357–359.

Turabian, K. L. 1955. A manual for writers of term papers, theses, and dissertations. Rev. ed., 82 p. Chicago: University of Chicago Press.

Ulman, J. N., Jr. 1952. Technical reporting. 289 p. New York: Henry Holt and Co.

U. S. Government Printing Office. 1945. Style manual. 435 p. Washington: Government Printing Office.

Walker, H. M., and J. Lev. 1953. Statistical inference. 507 p. New York: Henry Holt and Co.

Webster's New Collegiate Dictionary. 1949. 2nd ed., 1209 p. Springfield, Mass.: G. and C. Merriam Co.

Wilson, E. B., Jr. 1952. An introduction to scientific research. 375 p. New York: McGraw-Hill Book Co.

Wistar Institute. 1934. The Wistar Institute style brief. 169 p. Philadelphia: Wistar Institute Press.

Woodger, J. H. 1952. Biology and language. 364 p. Cambridge, England: Cambridge University Press.

Woolley, E. C. 1909. The mechanics of writing. 396 p. Boston: D. C. Heath and Co.

Woolley, E. C., F. W. Scott, and E. T. Berdahl. 1944. College handbook of composition. 4th ed., 452 p. Boston: D. C. Heath and Co.

Worthing, A. G., and J. Geffner. 1943. Treatment of experimental data. 342 p. New York: John Wiley and Sons.

Youden, W. J. 1951. Statistical methods for chemists. 126 p. New York: John Wiley and Sons.

Zirkle, C. 1955. Our splintered learning and the status of scientists. Science 121: 513–519.

INDEX

184 INDEX